LONDON MATHEMATICAL SOCIETY LECTURE NOTE SERIES

Managing Editor: Professor J.W.S. Cassels, Department of Pure Mathematics and Mathematical Statistics, University of Cambridge, 16 Mill Lane, Cambridge CB2 1SB, England

The books in the series listed below are available from booksellers, or, in case of difficulty, from Cambridge University Press.

34 Representation theory of Lie groups, M.F. ATIYAH *et al*
36 Homological group theory, C.T.C. WALL (ed)
39 Affine sets and affine groups, D.G. NORTHCOTT
40 Introduction to H_p spaces, P.J. KOOSIS
43 Graphs, codes and designs, P.J. CAMERON & J.H. VAN LINT
45 Recursion theory: its generalisations and applications, F.R. DRAKE & S.S. WAINER (eds)
46 p-adic analysis: a short course on recent work, N. KOBLITZ
49 Finite geometries and designs, P. CAMERON, J.W.P. HIRSCHFELD & D.R. HUGHES (eds)
50 Commutator calculus and groups of homotopy classes, H.J. BAUES
51 Synthetic differential geometry, A. KOCK
57 Techniques of geometric topology, R.A. FENN
59 Applicable differential geometry, M. CRAMPIN & F.A.E. PIRANI
62 Economics for mathematicians, J.W.S. CASSELS
66 Several complex variables and complex manifolds II, M.J. FIELD
69 Representation theory, I.M. GELFAND *et al*
74 Symmetric designs: an algebraic approach, E.S. LANDER
76 Spectral theory of linear differential operators and comparison algebras, H.O. CORDES
77 Isolated singular points on complete intersections, E.J.N. LOOIJENGA
78 A primer on Riemann surfaces, A.F. BEARDON
79 Probability, statistics and analysis, J.F.C. KINGMAN & G.E.H. REUTER (eds)
80 Introduction to the representation theory of compact and locally compact groups, A. ROBERT
81 Skew fields, P.K. DRAXL
82 Surveys in combinatorics, E.K. LLOYD (ed)
83 Homogeneous structures on Riemannian manifolds, F. TRICERRI & L. VANHECKE
86 Topological topics, I.M. JAMES (ed)
87 Surveys in set theory, A.R.D. MATHIAS (ed)
88 FPF ring theory, C. FAITH & S. PAGE
89 An F-space sampler, N.J. KALTON, N.T. PECK & J.W. ROBERTS
90 Polytopes and symmetry, S.A. ROBERTSON
91 Classgroups of group rings, M.J. TAYLOR
92 Representation of rings over skew fields, A.H. SCHOFIELD
93 Aspects of topology, I.M. JAMES & E.H. KRONHEIMER (eds)
94 Representations of general linear groups, G.D. JAMES
95 Low-dimensional topology 1982, R.A. FENN (ed)
96 Diophantine equations over function fields, R.C. MASON
97 Varieties of constructive mathematics, D.S. BRIDGES & F. RICHMAN
98 Localization in Noetherian rings, A.V. JATEGAONKAR
99 Methods of differential geometry in algebraic topology, M. KAROUBI & C. LERUSTE
100 Stopping time techniques for analysts and probabilists, L. EGGHE
101 Groups and geometry, ROGER C. LYNDON
103 Surveys in combinatorics 1985, I. ANDERSON (ed)
104 Elliptic structures on 3-manifolds, C.B. THOMAS
105 A local spectral theory for closed operators, I. ERDELYI & WANG SHENGWANG
106 Syzygies, E.G. EVANS & P. GRIFFITH

107 Compactification of Siegel moduli schemes, C-L. CHAI
108 Some topics in graph theory, H.P. YAP
109 Diophantine analysis, J. LOXTON & A. VAN DER POORTEN (eds)
110 An introduction to surreal numbers, H. GONSHOR
111 Analytical and geometric aspects of hyperbolic space, D.B.A.EPSTEIN (ed)
112 Low-dimensional topology and Kleinian groups, D.B.A. EPSTEIN (ed)
113 Lectures on the asymptotic theory of ideals, D. REES
114 Lectures on Bochner-Riesz means, K.M. DAVIS & Y-C. CHANG
115 An introduction to independence for analysts, H.G. DALES & W.H. WOODIN
116 Representations of algebras, P.J. WEBB (ed)
117 Homotopy theory, E. REES & J.D.S. JONES (eds)
118 Skew linear groups, M. SHIRVANI & B. WEHRFRITZ
119 Triangulated categories in the representation theory of finite-dimensional algebras, D. HAPPEL
121 Proceedings of *Groups - St Andrews 1985*, E. ROBERTSON & C. CAMPBELL (eds)
122 Non-classical continuum mechanics, R.J. KNOPS & A.A. LACEY (eds)
124 Lie groupoids and Lie algebroids in differential geometry, K. MACKENZIE
125 Commutator theory for congruence modular varieties, R. FREESE & R. MCKENZIE
126 Van der Corput's method of exponential sums, S.W. GRAHAM & G. KOLESNIK
127 New directions in dynamical systems, T.J. BEDFORD & J.W. SWIFT (eds)
128 Descriptive set theory and the structure of sets of uniqueness, A.S. KECHRIS & A. LOUVEAU
129 The subgroup structure of the finite classical groups, P.B. KLEIDMAN & M.W.LIEBECK
130 Model theory and modules, M. PREST
131 Algebraic, extremal & metric combinatorics, M-M. DEZA, P. FRANKL & I.G. ROSENBERG (eds)
132 Whitehead groups of finite groups, ROBERT OLIVER
133 Linear algebraic monoids, MOHAN S. PUTCHA
134 Number theory and dynamical systems, M. DODSON & J. VICKERS (eds)
135 Operator algebras and applications, 1, D. EVANS & M. TAKESAKI (eds)
136 Operator algebras and applications, 2, D. EVANS & M. TAKESAKI (eds)
137 Analysis at Urbana, I, E. BERKSON, T. PECK, & J. UHL (eds)
138 Analysis at Urbana, II, E. BERKSON, T. PECK, & J. UHL (eds)
139 Advances in homotopy theory, S. SALAMON, B. STEER & W. SUTHERLAND (eds)
140 Geometric aspects of Banach spaces, E.M. PEINADOR and A. RODES (eds)
141 Surveys in combinatorics 1989, J. SIEMONS (ed)
142 The geometry of jet bundles, D.J. SAUNDERS
143 The ergodic theory of discrete groups, PETER J. NICHOLLS
144 Introduction to uniform spaces, I.M. JAMES
145 Homological questions in local algebra, JAN R. STROOKER
146 Cohen-Macaulay modules over Cohen-Macaulay rings, Y. YOSHINO
147 Continuous and discrete modules, S.H. MOHAMED & B.J. MÜLLER
148 Helices and vector bundles, A.N. RUDAKOV et al
149 Solitons, nonlinear evolution equations and inverse scattering, M.A. ABLOWITZ & P.A. CLARKSON
150 Geometry of low-dimensional manifolds 1, S. DONALDSON & C.B. THOMAS (eds)
151 Geometry of low-dimensional manifolds 2, S. DONALDSON & C.B. THOMAS (eds)
152 Oligomorphic permutation groups, P. CAMERON
153 L-functions in arithmetic, J. COATES & M.J. TAYLOR (eds)
154 Number theory and cryptography, J. LOXTON (ed)
155 Classification theories of polarized varieties, TAKAO FUJITA
156 Twistors in mathematics and physics, T.N. BAILEY & R.J. BASTON (eds)
158 Geometry of Banach spaces, P.F.X. MÜLLER & W. SCHACHERMAYER (eds)

London Mathematical Society Lecture Note Series. 126

Van der Corput's Method of Exponential Sums

S. W. Graham
Michigan Technological University, USA
and
G. Kolesnik
California State University, Los Angeles, USA

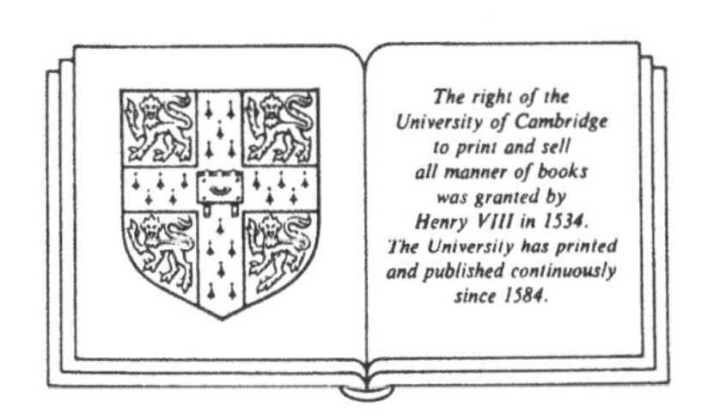

CAMBRIDGE UNIVERSITY PRESS
Cambridge
New York Port Chester Melbourne Sydney

CAMBRIDGE UNIVERSITY PRESS
Cambridge, New York, Melbourne, Madrid, Cape Town, Singapore, São Paulo

Cambridge University Press
The Edinburgh Building, Cambridge CB2 8RU, UK

Published in the United States of America by Cambridge University Press, New York

www.cambridge.org
Information on this title: www.cambridge.org/9780521339278

First published 1991
Re-issued in this digitally printed version 2008

A catalogue record for this publication is available from the British Library

ISBN 978-0-521-33927-8 paperback

TABLE OF CONTENTS

Acknowledgments

Several people read the preliminary drafts of this book, corrected errors, and made suggestions. It is our pleasure to thank Michael Filaseta, Mary Graham, Roger Heath-Brown, Martin Huxley, Matti Jutila, and Jeff Vaaler for their assistance. We also thank Hugh Montgomery for allowing us to use some unpublished material. We thank our editor, David Tranah, for his assistance and unflagging patience. Finally, the first author would like to thank Julian Gevirtz for introducing him to the world of electronic typesetting.

1. INTRODUCTION

1.1 BASIC DEFINITIONS

In this monograph, we will give an account of van der Corput's method in one and two dimensions. The purpose of this method is to obtain bounds for exponential sums, particularly those exponential sums that arise in number-theoretic problems.

An exponential sum is a sum of the form

$$\sum_{a<n\leq b} e(f(n)), \tag{1.1.1}$$

where $f(n)$ is a real-valued function, and the notation $e(x)$ is used to denote $e^{2\pi i x}$. Such sums arise in several different contexts in number theory. One of the simplest is in estimates for the Riemann-zeta function. The problem of obtaining an upper bound for $\zeta(\sigma+it)$ can be reduced to the problem of bounding

$$\sum_{N<n\leq 2N} n^{-\sigma+it}; \tag{1.1.2}$$

the details of the reduction will be given in Section 2.5. The $n^{-\sigma}$ can be removed via partial summation, and we are then left with a sum of the (1.1.1) with $f(n) = -\frac{t}{2\pi}\log n$.

Another simple example where exponential sums arise is in the Dirichlet divisor problem. Let $d(n)$ denote the numbers of divisors of the natural number n, and let

$$\Delta(x) = \sum_{n\leq x} d(n) - x\log x - (2\gamma-1)x, \tag{1.1.3}$$

where γ is Euler's constant. An elementary argument (see, for example, Chapter 18.2 of Hardy and Wright (1979)) can be used to show that $\Delta(x) \ll x^{1/2}$. Voronoï(1903) showed that the error term could be improved to $O(x^{1/3})$. As we shall see in Chapter 4,

$$\Delta(x) = -2\sum_{n\leq x^{1/2}} \psi(x/n) + O(1), \tag{1.1.4}$$

where $\psi(w) = w - [w] - 1/2$. Now ψ is a periodic function of period 1, and it can be expressed as a Fourier series. This gives a way of writing (1.1.4) in terms of exponential sums.

There are other problems that lead to consideration of sums of the form $\sum_n \psi(f(n))$, and we will discuss some of these in Chapter 4.

1.2 HISTORICAL OVERVIEW

The application of exponential sums to number theory began with the Weyl's paper, (Weyl 1916) "Über die Gleichverteilung von Zahlen mod. Eins." One of several new ideas that Weyl introduced was a useful transformation that arises upon squaring an exponential sum. Let S denote the sum in (1.1.1). Then

$$\begin{aligned}|S|^2 &= \sum_{a<m,n\le b} e(f(m)-f(n)) \\ &= \sum_{|h|<b-a} \sum_{n\in I_h} e(f(n+h)-f(n)) \end{aligned} \tag{1.2.1}$$

where $I_r = \{n : a < n, n+r \le b\}$. This is useful because the differenced function $f(n+r)-f(n)$ occurring in the inner sum is easier to handle than the original function $f(n)$. For example, if $f(n)$ is a polynomial of degree k, then $f(n+r)-f(n)$ is a polynomial of degree $k-1$. After $k-1$ applications of (1.2.1), the problem reduces to consideration of exponential sums of linear functions. The latter are easily handled because they are geometric series.

Van der Corput (1922) modified and improved Weyl's method as follows. Let H be an arbitrary positive integer. Then

$$HS = \sum_{h=1}^{H} \sum_{a-h<n\le b-h} e(f(n+h)).$$

By squaring both sides of the above and applying Cauchy's inequality, one obtains

$$|S|^2 \le \frac{|b-a|+H}{H} \sum_{|h|<H} \left(1-\frac{|h|}{H}\right) \sum_{n\in I_h} e(f(n+h)-f(n)). \tag{1.2.2}$$

The details of the derivation of (1.2.2) will be given in Section 2.3. Note that (1.2.2) is very similar to (1.2.1) when $H = b-a$.

Another innovation of van der Corput was a combination of the Poisson summation formula and the method of stationary phase. Under suitable conditions on f, it can be shown that

$$\sum_{n\in I} e(f(n)) = \sum_{\alpha\le\nu\le\beta} \frac{e(-\phi(\nu)-1/8)}{|f''(x_\nu)|^{1/2}} + \text{ error terms}, \tag{1.2.3}$$

where α and β are defined in terms of f', x_ν is defined by the relation $f'(x_\nu) = \nu$, and $\phi(\nu) = -f(x_\nu) + \nu x_\nu$. A more complete discussion of (1.2.3) will be given in Chapter 3.

Van der Corput used his method to prove that

$$\Delta(x) \ll x^{33/100}. \tag{1.2.4}$$

(Actually he proved (1.2.4) with the smaller exponent 163/494.) Walfisz (1924) used the same method to show that

$$\zeta(1/2+it) \ll t^{163/998}. \tag{1.2.5}$$

Van der Corput (1928) improved the exponent in the divisor problem to 27/82. Titchmarsh (1931) sharpened the exponent in (1.2.5) to 27/164, and Phillips (1933) further reduced the exponent to 229/1392.

A more important aspect of Phillips' 1933 paper was his refinement of exponent systems. Van der Corput (1922) defined an exponent system to be a set of ordered pairs $\{(k_1, l_1), \ldots, (k_r, l_r)\}$ such that, if $|f'| \approx y$ and $N_1 \leq 2N$ then

$$\sum_{N<n\leq N_1} e(f(n)) \ll y^{k_1}N^{l_1} + \ldots + y^{k_r}N^{l_r},$$

provided f satisfies a certain set of hypotheses. (A complete statement of the hypotheses is lengthy; see the discussion in Chapter 3.) Phillips gave a form of this theory in which only one pair (k, l) is required. Thus he refers to exponent pairs, rather than exponent systems. The trivial bound $\sum_{N<n\leq N_1} e(f(n)) \ll N$ shows that $(0, 1)$ is an exponent pair. Phillips showed that if (k, l) is an exponent pair then so are

$$A(k,l) = \left(\frac{k}{2k+2}, \frac{k+l+1}{2k+2}\right) \text{ and } B(k,l) = (l-1/2, k+1/2).$$

The A-result follows from (1.2.2), and the B-result follows from (1.2.3). The theory of exponent pairs will be discussed in Chapter 3.

It can be shown that if (k, l) is an exponent pair and $\theta(k,l) = (k+l-1/2)/2$, then

$$\zeta(1/2+it) \ll t^{\theta(k,l)} \log t.$$

For example, Phillips' exponent 229/1392 is $\theta(k, l)$ when

$$(k,l) = ABA^3BA^2BA^2B(0,1) = (97/696, 480/696).$$

This naturally leads to the question of finding the minimum of $\theta(k, l)$ for all exponent pairs that are obtainable from $(0, 1)$ by application of the A and B processes. Rankin (1955) found this minimum. Graham (1985) gave further details and considered the corresponding problem when θ is a rational function of k and l. Graham's method will be presented in Chapter 5.

1.3 TWO DIMENSIONAL SUMS

In some applications, sums of the form

$$\sum_{(m,n)\in \mathbf{D}} e(f(m,n))$$

arise, where $\mathbf{D}$ is some simple region in $\mathbf{R}^2$. These sums may be reduced to one-dimensional sums by noting that

$$\Big|\sum_{(m,n)\in\mathbf{D}} e(f(m,n))\Big| \le \sum_m \Big|\sum_{\substack{n\\(m,n)\in\mathbf{D}}} e(f(m,n))\Big|,$$

but such a simplification can be wasteful. A more efficient approach is to develop two-dimensional analogs of Process A and Process B. The first steps in this direction were taken by Titchmarsh (1934a) in an application to Epstein's zeta-function. Other authors extended the two dimensional method to bounds for $\zeta(1/2+it)$ and $\Delta(x)$. The best results obtained by this method to date are due to Kolesnik (1985), who shows that

$$\zeta(1/2+it) \ll t^{139/858+\epsilon}$$

and

$$\Delta(x) \ll x^{139/429+\epsilon}. \tag{1.3.1}$$

In Chapter 6, we give an overview of two-dimensional sums. The algorithm of Chapter 5 can be easily adapted to show the limitations of what can be accomplished with these sums. We will show for example, that with a two-dimensional method, the exponent in (1.3.1) cannot be made smaller than

$$0.32392\ldots.$$

(Note that $139/429 = 0.324009\ldots$.) We also prove two-dimensional analogs of the B and AB theorems.

1.4 THE METHOD OF BOMBIERI AND IWANIEC

The latest advance in estimation of exponential sums began with the work of Bombieri and Iwaniec (1985a,b). They proved that for any $\epsilon > 0$,

$$\zeta(1/2+it) \ll t^{9/56+\epsilon}.$$

This is better than any result that can be obtained from either the one-dimensional or two-dimensional methods. Their work uses the Weyl-van der Corput inequality (1.2.2) and the method of stationary phase. They also use an elaborate form of the large sieve inequality and some mean value theorems that are reminiscent of Vinogradov's method.

Huxley and Watt (1988) generalize the Bombieri-Iwaniec result to show that for every $\epsilon > 0$,

$$\left(\frac{9}{56}+\epsilon, \frac{37}{56}+\epsilon\right)$$

is an exponent pair. We will prove this result in Chapter 7.

The line of research opened up by Bombieri and Iwaniec has proved promising, and we expect further progress along the same lines in the future. Iwaniec and Mozzochi (1988) proved that

$$\Delta(x) \ll x^{7/22+\epsilon}.$$

Watt (1989) has shown that

$$\zeta(1/2+it) \ll t^{89/560+\epsilon}.$$

1.5 NOTATION

As mentioned earlier, $e(x)$ denotes $e^{2\pi ix}$, $[x]$ denotes the greatest integer $\leq x$, and $\psi(x) = x - [x] - 1/2$.

We will make extensive use of the Vinogradov notations $\ll$ and $\gg$. For functions f and g with g taking non-negative real values, $f \ll g$ means $|f| \leq C|g|$ for some unspecified constant C. If f is also non-negative, $f \gg g$ means $g \ll f$. We also use the Landau notation $f = O(g)$; this is equivalent to $f \ll g$. The notation $f \approx g$ means that $f \ll g$ and $g \ll f$.

The notation $||x||$ denotes the distance to the nearest integer of x; i.e.

$$||x|| = \min_{n \in \mathbf{Z}} |x-n|.$$

We also use the generalized factorial $(\alpha)_n$. This is defined for non-negative integers n by the recursive relations

$$(\alpha)_0 = 1\,,\ (\alpha)_{n+1} = (\alpha+n)(\alpha)_n.$$

2. THE SIMPLEST VAN DER CORPUT ESTIMATES

2.1 ESTIMATES USING FIRST AND SECOND DERIVATIVES

In the next two chapters, we will obtain upper bounds for sums of the form

$$S = \sum_{n \in I} e(f(n)), \tag{2.1.1}$$

where $I = (a, b]$. (We assume throughout this chapter that a and b are integers.) In this chapter, we obtain the simplest such upper bounds.

From the triangle inequality, we see that

$$|S| \leq \sum_{n \in I} |e(f(n)| = |I| = b - a. \tag{2.1.2}$$

This estimate (henceforth called the *trivial estimate*) holds with equality if $f(n) = xn + y$ and x is an integer. Thus any improvement on (2.1.2) will require some hypothesis on f.

It is instructive to examine the case $f(n) = xn + y$ more closely. If x is not an integer, then

$$|\sum_{n \in I} e(xn + y)| = \frac{|e(bx) - e(ax)|}{|e(x) - 1|} \leq \frac{1}{|\sin \pi x|}. \tag{2.1.3}$$

An examination of the graphs $y = \sin \pi x$ and $y = 2x$ shows that $\sin \pi x \geq 2x$ for $0 \leq x \leq 1/2$. Now define

$$||x|| = \min\{|x - n| : n \in \mathbf{Z}\}.$$

Since $|\sin \pi x|$ is periodic with period 1 and $\sin \pi(1 - x) = \sin \pi x$, we have

$$|\sin \pi x| \geq 2||x||$$

for all real x. Consequently,

$$|\sum_{n \in I} e(xn + y)| \ll ||x||^{-1}.$$

Our first theorem generalizes this result.

THEOREM 2.1. *(Kusmin-Landau) If f is continuously differentiable, f' is monotonic, and $\|f'\| \geq \lambda > 0$ on I then*

$$\sum_{n \in I} e(f(n)) \ll \lambda^{-1}.$$

Proof. Since

$$\Big|\sum_{n \in I} e(f(n))\Big| = \Big|\sum_{n \in I} e(-f(n))\Big|,$$

we may assume that f' is increasing. Our hypotheses imply that for some integer k,

$$k + \lambda \leq f'(n) \leq k + 1 - \lambda.$$

Now

$$\sum_{n \in I} e(f(n)) = \sum_{n \in I} e(f(n) - kn),$$

so we may assume that $\lambda \leq f'(n) \leq 1 - \lambda$.

Define $g(n) = f(n+1) - f(n)$. From the mean value theorem, we see that there is some x_n such that $n \leq x_n \leq n+1$ and $g(n) = f'(x_n)$. Consequently, g is increasing and $\lambda \leq g(n) \leq 1 - \lambda$. We write

$$e(f(n)) = \frac{e(f(n)) - e(f(n+1))}{1 - e(g(n))} = \{e(f(n)) - e(f(n+1))\}C_n,$$

where $C_n = \frac{1}{2}(1 + i \cot \pi g(n))$. Therefore

$$\begin{aligned}\sum_{n \in I} e(f(n)) &= \sum_{n=a+1}^{b-1} \{e(f(n)) - e(f(n+1)\}C_n + e(f(b)) \\ &= \sum_{n=a+2}^{b-1} e(f(n))(C_n - C_{n-1}) + e(f(a+1))C_{a+1} + e(f(b))(1 - C_{b-1}),\end{aligned}$$

and thus

$$\Big|\sum_{n \in I} e(f(n))\Big| \leq \frac{1}{2} \sum_{n=a+2}^{b-1} |\cot(\pi g(n-1)) - \cot(\pi g(n))| + |C_{a+1}| + |1 - C_{b-1}|.$$

Since $\cot \pi g(n)$ is a decreasing function, the absolute value bars in the right-hand sum may be removed. The remaining sum telescopes, so

$$\Big|\sum_{n \in I} e(f(n))\Big| \leq \frac{1}{2}\{\cot(\pi g(a+1)) - \cot(\pi g(b-1))\} + |C_{a+1}| + |1 - C_{b-1}|.$$

The theorem follows by using the bound $|\cot \pi x| \ll \|x\|^{-1}$.

It is clear from (2.1.3) that Theorem 2.1 is sharp. On the other hand, the hypothesis $\|f'\| \geq \lambda$ is rather restrictive. Our next theorem relaxes this hypothesis.

THEOREM 2.2. *(van der Corput) Suppose that f is a real valued function with two continuous derivatives on I. Suppose also that there is some $\lambda > 0$ and some $\alpha \geq 1$ such that*

$$\lambda \leq |f''(x)| \leq \alpha\lambda$$

on I. Then

$$\sum_{n\in I} e(f(n)) \ll \alpha|I|\lambda^{1/2} + \lambda^{-1/2}.$$

Proof. Let $\delta(< 1/2)$ be a parameter to be chosen later. From the hypothesis, it follows that I can be split into $\leq \alpha|I|\lambda+2$ intervals on which $||f'|| \geq \delta$, and $\leq \alpha|I|\lambda+1$ other intervals, each of length $\leq 2\delta/\lambda$. We apply Theorem 2.1 to the former set of intervals, and the trivial estimate to the latter. We get

$$\sum_{n\in I} e(f(n)) \ll (\alpha|I|\lambda + 1)(1/\delta + \delta/\lambda + 1).$$

We choose $\delta = \lambda^{1/2}$; this proves the desired result if $\lambda \leq 1/4$. If $\lambda > 1/4$, the result follows from the trivial estimate.

2.2 SOME SIMPLE INEQUALITIES

In this section, we will derive a couple of results that we shall be using throughout this book. Both are simple but, nonetheless, useful.

LEMMA 2.3. *If $X_1, \ldots, X_k$ are positive numbers and $a_1, \ldots, a_k$ are non-negative numbers such that $a_1 + \ldots + a_k = 1$, then*

$$\min(X_1, \ldots, X_k) \leq X_1^{a_1} \ldots X_k^{a_k} \leq \max(X_1, \ldots, X_k).$$

The proof is obvious. We shall often refer to this as the *convexity principle.*

Our second Lemma generalizes the following well known principle. Suppose we have an estimate of the form

$$U \ll AH^a + BH^{-b},$$

where A, B, a, and b are positive constants and H is at our disposal. We obviously want to take H so as to minimize the right hand side. By choosing H to satisfy $AH^a = BH^{-b}$, we get

$$U \ll (A^b B^a)^{1/(a+b)},$$

and this is best possible apart from the value of the implied constant. To generalize this principle, we prove

LEMMA 2.4. *Suppose that*

$$L(H) = \sum_{i=1}^{m} A_i H^{a_i} + \sum_{j=1}^{n} B_j H^{-b_j},$$

where A_i, B_j, a_i, and b_j are positive. Assume that $H_1 \le H_2$. Then there is some H with $H_1 \le H \le H_2$ and

$$L(H) \ll \sum_{i=1}^{m} \sum_{j=1}^{n} (A_i^{b_j} B_j^{a_i})^{1/(a_i+b_j)} + \sum_{i=1}^{m} A_i H_1^{a_i} + \sum_{j=1}^{n} B_j H_2^{-b_j}.$$

The implied constants depend only on m and n.

Proof. Define

$$L_+(H) = \max(A_1 H^{a_1}, \ldots, A_m H^{a_m})$$

and

$$L_-(H) = \max(B_1 H^{-b_1}, \ldots, B_n H^{-b_n}).$$

Now $L \le mL_+ + nL_-$, so it suffices to bound L_+ and L_-. We observe that L_+ is a strictly increasing continuous function, $L_+(0) = 0$, and $L_+(\infty) = \infty$. Similarly L_- is a strictly decreasing continuous function, $L_-(0) = \infty$, and $L_-(\infty) = 0$. Therefore there is a unique H_0 such that $L_+(H_0) = L_-(H_0)$. We distinguish three cases: (a) $H_1 \le H_0 \le H_2$, (b) $H_0 < H_1$, (c) $H_0 > H_2$.

If $H_1 \le H_0 \le H_2$ then there is some i and some j such that

$$L_+(H_0) = A_i H_0^{a_i} = L_-(H_0) = B_j H_0^{-b_j}.$$

Consequently

$$L_+(H_0) = L_-(H_0) = (A_i^{b_j} B_j^{a_i})^{1/(a_i+b_j)}.$$

If $H_0 < H_1$ then

$$L_-(H_1) < L_+(H_1) \le \sum_{i=1}^{m} A_i H_1^{a_i}.$$

If $H_0 > H_2$ then

$$L_+(H_2) < L_-(H_2) \le \sum_{j=1}^{n} B_j H_2^{-b_j}.$$

2.3 THE WEYL-VAN DER CORPUT INEQUALITY

Theorem 2.2 has the disadvantage of being worse than the trivial estimate if $\lambda > 1$. This limitation can be overcome if we replace f by a function with smaller derivatives; the next lemma allows us to make such a replacement.

LEMMA 2.5. *(Weyl-van der Corput) Suppose $\xi(n)$ is a complex valued function such that $\xi(n) = 0$ if $n \notin I$. If H is a positive integer then*

$$|\sum_n \xi(n)|^2 \leq \frac{|I|+H}{H} \sum_{|h|<H} \left(1 - \frac{|h|}{H}\right) \sum_n \xi(n)\overline{\xi(n-h)}.$$

Proof. Observe that

$$H \sum_n \xi(n) = \sum_{k=1}^{H} \sum_n \xi(n+k) = \sum_n \sum_{k=1}^{H} \xi(n+k).$$

The inner sum is empty unless $a - H < n \leq b - 1$. By Cauchy's inequality,

$$\begin{aligned} H^2 |\sum_n \xi(n)|^2 &\leq (|I| + H) \sum_n |\sum_{k=1}^{H} \xi(n+k)|^2 \\ &= (|I| + H) \sum_{k=1}^{H} \sum_{l=1}^{H} \sum_n \overline{\xi(n+k)}\xi(n+l). \end{aligned}$$

We collect together the terms with $l - k = h$ to get the result.

For applications of Lemma 2.5 in this chapter, we will take

$$\xi(n) = \begin{cases} e(f(n)) & \text{if } n \in I; \\ 0 & \text{otherwise.} \end{cases}$$

Let S be as in (2.1.1) and define

$$S_1(h) = \sum_{n \in I(h)} e(f(n+h) - f(n)), \tag{2.3.1}$$

where $I(h) = \{n : n \in I \text{ and } n + h \in I\}$. We see that

$$|S|^2 \leq \frac{(|I| + H)}{H} \sum_{|h|<H} |S_1(h)|.$$

If we also assume that $H \leq |I|$ then

$$|S|^2 \leq \frac{2|I|}{H} \sum_{|h|<H} |S_1(h)|. \tag{2.3.2}$$

Up to now, we have assumed that H is an integer. If we allow H to be an arbitrary positive number and set $H_0 = [H] + 1$, we get

$$|S|^2 \le \frac{2|I|}{H_0} \sum_{|h|<H_0} |S_1(h)| \le \frac{2|I|}{H} \sum_{|h|\le H} |S_1(h)|. \tag{2.3.3}$$

Furthermore, if $h > 0$ then

$$S_1(h) = \sum_{a<n\le b-h} e(f(n+h) - f(n))$$

and

$$S_1(-h) = \sum_{a+h<n\le b} e(f(n-h) - f(n)).$$

By making a change of variable, we see that $S_1(-h) = \overline{S_1(h)}$. Therefore

$$|S|^2 \le \frac{2|I|^2}{H} + \frac{4|I|}{H} \sum_{1\le h\le H} |S_1(h)|. \tag{2.3.4}$$

We make two further comments that will be relevant in the next section. Suppose I' is an interval that contains I. Then the proof of (2.3.4) may be easily modified to yield

$$|S|^2 \le \frac{2|I'|^2}{H} + \frac{4|I'|}{H} \sum_{1\le h\le H} |S_1(h)|. \tag{2.3.5}$$

provided only that $H \le |I'|$. There are times (cf. Theorem 2.8) when this weaker condition on H is convenient. The second comment is that the function $f(n+h)-f(n)$ may be written as

$$\int_0^1 \frac{\partial}{\partial t} f(n+ht)\, dt.$$

Such integral representations are convenient when Lemma 2.5 is iterated.

By combining (2.3.4) and Theorem 2.2, we prove

THEOREM 2.6. *Suppose that f is a real valued function with three continuous derivatives on I. Suppose also that for some $\lambda > 0$ and for some $\alpha \ge 1$,*

$$\lambda \le |f^{(3)}(x)| \le \alpha\lambda$$

on I. Then

$$S \ll |I|\lambda^{1/6}\alpha^{1/3} + |I|^{3/4}\alpha^{1/4} + |I|^{1/4}\lambda^{-1/4}.$$

Proof. It suffices to bound

$$\frac{2|I|^2}{H} + \frac{4|I|}{H} \sum_{1\le h\le H} |S_1(h)|.$$

If we set $f_1(n;h) = f(n+h) - f(n)$, then

$$f_1''(n;h) = \int_0^1 hf^{(3)}(n+th)\, dt,$$

and so

$$h\lambda \le |f_1''(n;h)| \le \alpha h\lambda.$$

From Theorem 2.2 we see that

$$|S_1(h)| \ll |I|h^{1/2}\alpha\lambda^{1/2} + h^{-1/2}\lambda^{-1/2}.$$

Consequently,

$$|S|^2 \ll |I|^2H^{-1} + |I|^2H^{1/2}\lambda^{1/2}\alpha + |I|H^{-1/2}\lambda^{-1/2}.$$

We apply Lemma 2.4 with $H_2 = |I|, H_1 = 0$ to get

$$|S|^2 \ll |I|^2\lambda^{1/3}\alpha^{2/3} + |I|^{3/2}\alpha^{1/2} + |I| + |I|^{1/2}\lambda^{-1/2}.$$

Because the second term dominates the third, the result follows.

We now give an example which provides a preview of the next section. We shall obtain upper bounds for

$$S = \sum_{N<n\le 2N} e(An^{-\sigma}),$$

where σ is a real number, $\sigma \ne 0, -1$, or -2, and $A > 0$.

In this example, all implied constants are allowed to depend on σ, but they are otherwise absolute. Let $f(n) = An^{-\sigma}$ and $F = AN^{-\sigma}$. By Theorems 2.2 and 2.6,

$$S \ll F^{1/2} + F^{-1/2}N \tag{2.3.6}$$

and

$$S \ll F^{1/6}N^{1/2} + N^{3/4} + F^{-1/4}N. \tag{2.3.7}$$

In fact, these estimates may be sharpened to

$$S \ll F^{1/2} + F^{-1}N \tag{2.3.8}$$

and

$$S \ll F^{1/6}N^{1/2} + F^{-1}N. \tag{2.3.9}$$

To prove (2.3.8), assume first that $\sigma > -1$. If $F \ge (2|\sigma|)^{-1}N$, then $S \ll F^{1/2}$ by (2.3.6). In the contrary case, we can apply the Kusmin-Landau inequality to get $S \ll F^{-1}N$.

If $\sigma \le -1$, we argue in a similar fashion; we apply (2.3.6) if $F \ge 2^{\sigma}|\sigma|^{-1}N$ and Kusmin-Landau otherwise.

To prove (2.3.9), we let

$$E_1 = \max(F^{1/2}, F^{-1}N) \text{ and } E_2 = \max(F^{1/6}N^{1/2}, N^{3/4}, F^{-1/4}N).$$

Therefore, $S \ll \min(E_1, E_2)$ and the desired estimate follows immediately unless $E_1 = F^{1/2}$ and $E_2 = N^{3/4}$ or $F^{-1/4}N$. If the first possibility for E_2 holds, then we use Lemma 2.3 (convexity) to get

$$S \ll E_1^{1/3} E_2^{2/3} = F^{1/6}N^{1/2}.$$

Otherwise, we use the same Lemma to get

$$S \ll E_1^{5/9} E_2^{4/9} = F^{1/6}N^{4/9} \leq F^{1/6}N^{1/2}.$$

2.4 ITERATING WEYL-VAN DER CORPUT

We begin by squaring (2.3.4) and using Cauchy's inequality to get

$$|S|^4 \leq \frac{8|I|^4}{H_1^2} + \frac{32|I|^2}{H_1} \sum_{1 \leq h_1 \leq H} |S_1(h_1)|^2.$$

By applying (2.3.5) to $S_1(h_1)$, we get

$$|S|^4 \leq \frac{8|I|^4}{H_1^2} + \frac{64|I|^4}{H_2} + \frac{128|I|^3}{H_1 H_2} \sum_{1 \leq h_1 \leq H_1} \sum_{1 \leq h_2 \leq H_2} |S_2(h_1, h_2)|,$$

provided $H_1 \leq |I|$ and $H_2 \leq |I|$. The summands on the right are defined by

$$S_2(h_1, h_2) = \sum_{n \in I(h_1, h_2)} e(f_2(n; h_1, h_2)),$$

$$\begin{aligned} f_2(n; h_1, h_2) &= f(n + h_1 + h_2) - f(n + h_1) - f(n + h_2) + f(n) \\ &= \int_0^1 \int_0^1 \frac{\partial^2}{\partial t_1 \partial t_2} f(n + t_1 h_1 + t_2 h_2)\, dt_1 dt_2, \end{aligned}$$

and

$$I(h_1, h_2) = \{n : n, n + h_1, n + h_2, \text{ and } n + h_1 + h_2 \in I\}.$$

If we assume that $H_1^2 = H_2$, then we may write

$$|S|^4 \leq 8^3 \Big\{ \frac{|I|^4}{H_2} + \frac{|I|^3}{H_1 H_2} \sum_{1 \leq h_1 \leq H_1} \sum_{1 \leq h_2 \leq H_2} |S_2(h_1, h_2)| \Big\}.$$

This argument may be continued, and a straightforward induction argument yields

LEMMA 2.7. *Let q be a positive integer, and let $Q = 2^q$. If $H \le |I|$, $H_q = H$, $H_{q-1} = H^{1/2}, \ldots,$ and $H_1 = H^{2/Q}$, then*

$$|S|^q \le 8^{Q-1}\left\{\frac{|I|^Q}{H} + \frac{|I|^{Q-1}}{H_1 \ldots H_q} \sum_{1 \le h_1 \le H_1} \cdots \sum_{1 \le h_q \le H_q} |S_q(\mathbf{h})|\right\},$$

where $\mathbf{h} = (h_1, h_2, \ldots h_q)$,

$$S_q(\mathbf{h}) = \sum_{n \in I(\mathbf{h})} e(f_q(n; \mathbf{h})),$$

$$f_q(n; \mathbf{h}) = \int_0^1 \cdots \int_0^1 \frac{\partial^q}{\partial t_1 \partial t_2 \ldots \partial t_q} f(n + \mathbf{h} \cdot \mathbf{t})\, dt_1 \ldots dt_q,$$

$$\mathbf{t} = (t_1, \ldots, t_q), \text{ and } I(\mathbf{h}) = (a, b - h_1 - h_2 - \ldots - h_q].$$

We combine Lemma 2.7 with Theorem 2.2 to get

THEOREM 2.8. *Let q be a positive integer. Suppose that f is a real valued function with $q+2$ continuous derivatives on I. Suppose also that for some $\lambda > 0$ and for some $\alpha \ge 1$,*

$$\lambda \le |f^{(q+2)}(x)| \le \alpha\lambda$$

on I. Let $Q = 2^q$. Then

$$S \ll |I|(\alpha^2\lambda)^{1/(4Q-2)} + |I|^{1-1/2Q}\alpha^{1/2Q} + |I|^{1-2/Q+1/Q^2}\lambda^{-1/2Q}.$$

The implied constant is absolute.

Proof. The function $f_q(n; \mathbf{h})$ of the previous lemma is equal to

$$\int_0^1 \cdots \int_0^1 h_1 \ldots h_q f^{(q)}(n + \mathbf{h} \cdot \mathbf{t})\, dt_1 \ldots dt_q;$$

therefore,

$$h_1 h_2 \ldots h_q \lambda \le |f_q''(n; \mathbf{h})| \le h_1 h_2 \ldots h_q \alpha\lambda.$$

By Theorem 2.2,

$$|S_q(\mathbf{h})| \ll |I|\alpha(h_1 h_2 \ldots h_q\lambda)^{1/2} + (h_1 h_2 \ldots h_q\lambda)^{-1/2}.$$

We now sum over $\mathbf{h}$. It is useful to note that

$$\sum_{1 \le h \le H} h^{-1/2} \le \sum_{1 \le h \le H} \int_{h-1}^{h} t^{-1/2}\, dt = \int_0^H t^{-1/2}\, dt = 2H^{1/2};$$

whence

$$(H_1 \dots H_q)^{-1} \sum_{\mathbf{h}} (h_1 \dots h_q)^{-1/2} \le 2^q (H_1 \dots H_q)^{-1/2}.$$

Similarly,

$$\sum_{1 \le h \le H} h^{1/2} \le H^{1/2} \sum_{1 \le h \le H} 1 \le H^{3/2},$$

and

$$(H_1 \dots H_q)^{-1} \sum_{\mathbf{h}} (h_1 \dots h_q)^{1/2} \le (H_1 \dots H_q)^{1/2}.$$

All together, we get

$$|S|^Q \ll (2^q 8^Q) |I|^Q (H^{-1} + H^{1-1/Q} \alpha \lambda^{1/2} + H^{-1+1/Q} |I|^{-1} \lambda^{-1/2}).$$

We apply Lemma 2.4 to get

$$|S|^Q \ll (2^q 8^Q) |I|^Q \{(\alpha^2 \lambda)^{Q/(4Q-2)} + |I|^{-1/2} \alpha^{1/2} + |I|^{-2+1/Q} \lambda^{-1/2}\}.$$

Taking Qth roots gives the result. Note that $(2^q 8^Q)^{1/Q}$ is bounded, so the implied constant is absolute.

We saw in an example in the previous section how an estimate obtained from Theorem 2.6 can be improved if there is sufficient information about f', f'' , and f'''. Our next theorem carries this idea further.

THEOREM 2.9. *Let $q \ge 0$ be an integer. Suppose that f has $q+2$ continuous derivatives on I, and that $I \subseteq (N, 2N]$. Assume also that there is some constant F such that*

$$|f^{(r)}(x)| \approx FN^{-r} \tag{2.4.1}$$

for $r = 1, \dots, q+2$. Then

$$S \ll F^{1/(4Q-2)} N^{1-(q+2)/(4Q-2)} + F^{-1} N.$$

The implied constant depends only upon the implied constants in (2.4.1).

Proof. By (2.4.1) when $r = 1$, we see that there are constants c_1 and c_2 such that

$$c_1 FN^{-1} \le |f'(n)| \le c_2 FN^{-1}.$$

If $c_2 FN^{-1} \le 1/2$, then we can use Theorem 2.1 (Kusmin-Landau) to get $S \ll F^{-1}N$. In the contrary case, Theorem 2.2 yields $S \ll F^{1/2}$. In any case,

$$S \ll F^{1/2} + F^{-1} N,$$

which is the desired result for $q = 0$.

We complete the proof by induction on q. Assume the theorem is true with q replaced by $q-1$. From the induction hypothesis and Theorem 2.8, we see that $S \ll \min(E_1, E_2)$,where

$$E_1 = \max(F^{1/(2Q-2)}N^{1-(q+1)/(2Q-2)}, F^{-1}N) = \max(M_1, M_2),$$

say, and

$$\begin{aligned} E_2 &= \max(F^{1/(4Q-2)}N^{1-(q+2)/(4Q-2)}, N^{1-1/2Q}, F^{-1/2Q}N^{1+(q-2)/2Q+1/Q^2}) \\ &= \max(M_3, M_4, M_5), \end{aligned}$$

say. The result follows immediately unless we have one of the following cases: (i) $E_1 = M_1, E_2 = M_4$, (ii) $E_1 = M_1, E_2 = M_5$.

In case (i), we get the result by using

$$\min(M_1, M_4) \le (M_1^{2Q-2}M_4^{2Q})^{1/(4Q-2)}.$$

In case (ii), we use

$$\min(M_1, M_5) \le M_1^a M_5^b,$$

where

$$a = \frac{(2Q-2)(6Q-2)}{(4Q-2)^2}, \quad b = \frac{4Q^2}{(4Q-2)^2},$$

and the result follows again.

2.5 UPPER BOUNDS FOR THE RIEMANN ZETA-FUNCTION

In this section, we will use $s = \sigma + it$ to denote a complex variable.

Riemann's zeta-function is defined as

$$\zeta(s) = \sum_{n=1}^{\infty} n^{-s}$$

for $\sigma > 1$. This function may be continued to function which is analytic in the whole plane except for a simple pole at $s = 1$. For our purposes, we need this continuation only for $\sigma > 0$; we do this in the course of proving Lemma 2.11.

Here we shall bound $\zeta(s)$ in the range $1/2 \le \sigma \le 1$ and $t \ge 3$. The latter restriction allows us to avoid the pole at $s = 1$ and to regard $\log t$ and $\log\log t$ as positive quantities.

Our first two lemmas reduce the problem to estimates of finite exponential sums.

LEMMA 2.10. *If $1/2 \le \sigma \le 1$ and $N \le M$ then*

$$\sum_{N<n\le M} n^{-\sigma-it} \ll N^{-\sigma} \max_{N<u\le M} \Big|\sum_{N<n\le u} n^{it}\Big|.$$

Proof. Let

$$S(u) = \sum_{N<n\le u} n^{-it}.$$

Then

$$\begin{aligned}\sum_{N<n\le M} n^{-\sigma-it} &= \int_N^M u^{-\sigma}\, dS(u) = S(M)M^{-\sigma} + \sigma \int_N^M S(u)u^{-\sigma-1}\, du \\ &\ll N^{-\sigma} \max_{N<u\le M} |S(u)|,\end{aligned}$$

and the desired result follows.

LEMMA 2.11. *If $1/2 \le \sigma \le 1$ and $t \ge 3$ then*

$$|\zeta(\sigma+it)| \ll \Big|\sum_{n\le t} n^{-\sigma-it}\Big| + t^{1-2\sigma}\log t.$$

Proof. If $\sigma > 1$ and $M \ge 1$ then

$$\zeta(s) = \sum_{n\le M} n^{-s} + \int_M^\infty u^{-s} d[u] = \sum_{n\le M} n^{-s} + \frac{M^{1-s}}{s-1} + s\int_M^\infty \frac{[u]-u}{u^{s+1}}\, du.$$

The last integral converges for $\sigma > 0$, so this gives an analytic continuation of $\zeta(s)$ to the region $\sigma > 0, s \ne 1$. We set $M = t^2$ and use the inequality $|[u]-u| \le 1$ to get

$$\zeta(s) = \sum_{n\le t^2} n^{-s} + O(t^{1-2\sigma}).$$

The sum over the range $t < n \le t^2$ may be divided into $\ll \log t$ subsums of the form

$$\sum_{N<n\le N_1} n^{-\sigma-it}.$$

where $N_1 = \min(2N, t^2)$. From Lemma 2.10 and Theorem 2.1 (Kusmin-Landau), we see that each of the above subsums is $\ll N^{1-\sigma}t^{-1} \ll t^{1-2\sigma}$, and the result follows.

THEOREM 2.12. *Let $q \ge 1$ be an integer, $Q = 2^q$, and $\sigma(q) = 1 - (q+2)/(4Q-2)$. If $t \ge 3$ then*

$$\zeta(\sigma(q)+it) \ll t^{1/(4Q-2)}\log t.$$

In particular,

$$\zeta(1/2+it) \ll t^{1/6}\log t,$$

and

$$\zeta(5/7+it) \ll t^{1/14}\log t.$$

Proof. By the previous two lemmas, it suffices to show that

$$\sum_{N<n\leq M} n^{it} \ll N^{\sigma(q)}t^{1/(4Q-2)}$$

whenever $M \leq 2N$ and $N \leq t$. This, however, follows immediately from Theorem 2.9 with $f(t) = -(t\log n)/2\pi$ and $F = t$.

Since we appealed to Theorem 2.9, our estimate in the previous theorem depends on q. We could, however, remove such dependence by arguing more carefully from Theorem 2.8. An example of such an argument appears in Theorem 2.14.

Our next result uses the convexity principle to "interpolate" between $\sigma(q)$ and $\sigma(q+1)$.

COROLLARY 2.13. *Let q, Q, and $\sigma(q)$ be as in the previous theorem. If $\sigma(q) \leq \sigma \leq \sigma(q+1)$ and $t \geq 3$ then*

$$\zeta(\sigma+it) \ll t^{\beta(\sigma)}\log t,$$

where

$$\beta(\sigma) = \frac{\sigma(q+1)-\sigma}{\sigma(q+1)-\sigma(q)}\cdot\frac{1}{4Q-2} + \frac{\sigma-\sigma(q)}{\sigma(q+1)-\sigma(q)}\cdot\frac{1}{8Q-2}.$$

Proof. If $M \leq 2N$ and $N \leq t$ then

$$\sum_{N<n\leq M} n^{-it} \ll \min(E_1, E_2),$$

where

$$E_1 = N^{\sigma(q)}t^{1/(4Q-2)}, E_2 = N^{\sigma(q+1)}t^{1/(8Q-2)}.$$

Now we use $\min(E_1,E_2) \ll E_1^a E_2^b$, where a and b are chosen so that $a+b=1$ and $a\sigma(q)+b\sigma(q+1)=\sigma$. In other words,

$$a = \frac{\sigma(q+1)-\sigma}{\sigma(q+1)-\sigma(q)},\ b = \frac{\sigma-\sigma(q)}{\sigma(q+1)-\sigma(q)}.$$

It follows that

$$\sum_{N<n\leq M} n^{it} \ll t^{\beta(\sigma)}N^{\sigma},$$

and the desired result follows from Lemmas 2.10 and 2.11.

THEOREM 2.14. *If $t \geq 3$ then*

$$\zeta(1+it) \ll \frac{\log t}{\log\log t}.$$

Proof. From Lemma 2.11, we have

$$\zeta(1+it) = \sum_{n \leq R} n^{-1-it} + \sum_{R<n\leq t} n^{-1-it} + O(t^{-1}\log t). \tag{2.5.1}$$

The first sum is trivially $\ll \log R$. Our approach is to make the second sum $\ll 1$ for R as large as possible.

Suppose $R \leq N \leq t$ and $M \leq 2N$. Let

$$S = \sum_{N<n\leq M} n^{-it} = \sum_{N<n\leq M} e(f(n)),$$

say. For $q \geq 1$, we have

$$|f^{(q+2)}(n)| = \frac{(q+1)!t}{2\pi n^{q+2}}.$$

Therefore, the hypotheses of Theorem 2.8 are satisfied with

$$\lambda = \frac{(q+1)!t}{2\pi(2N)^{q+2}}, \quad \alpha = 2^{q+2}.$$

Consequently,

$$S \ll t^{1/(4Q-2)}N^{1-(q+2)/(4Q-2)} + N^{1-1/2Q} + t^{-1/2Q}N^{1+(q-2)/2Q+1/Q^2},$$

and this estimate is uniform in q. We also have $S \ll t^{1/2}$ from Theorem 2.2. We can now repeat the argument of Theorem 2.9 to obtain

$$S \ll t^{1/(4Q-2)}N^{1-(q+2)/(4Q-2)}. \tag{2.5.2}$$

This estimate is valid for $q \geq 0$, and the implied constant does not depend on q.

We use a splitting argument, Lemma 2.10, and (2.5.2) to get

$$\sum_{R<n\leq t} n^{-1-it} \ll t^{1/(4Q-2)}R^{-(q+2)/(4Q-2)}\log t \tag{2.5.3}$$

for $q \geq 0$. To make an appropriate choice of q, we observe that the left hand side of (2.5.3) is

$$\exp\left(\log\log t + \frac{\log t - (q+2)\log R}{4Q-2}\right),$$

and this is $\ll 1$ if

$$(\log\log t)(4Q-2) + \log t \leq (q+2)\log R. \tag{2.5.4}$$

We now let $R = \exp(\log t / c \log\log t)$ and $q = [b \log t / \log R] - 2$, where b and c will be chosen later. Then

$$4Q = \exp((q+2)\log 2) \le \exp(b \log t \log 2/\log R) = (\log t)^{bc\log 2}.$$

We choose $b = 3/2$ and $c = 1/(2\log 2)$. Then $4Q \le (\log t)^{3/4}$, so

$$(\log\log t)(4Q-2) + \log t \le \log t + (\log\log t)(\log t)^{3/4} \le b \log t \le (q+2)\log R$$

for t sufficiently large. Thus (2.5.4) is satisfied. From (2.5.1) and our choice of R we get

$$\zeta(1+it) \ll \log R \ll \frac{\log t}{\log\log t}.$$

2.6 NOTES

Theorem 2.1, while sufficient for our purposes, is a weak version of what is usually referred to as the Kusmin-Landau inequality. Suppose f and θ are such that

$$0 < \theta \le f(2) - f(1) \le \ldots \le f(N) - f(N-1) < 1 - \theta$$

and $S = \sum_{n=1}^{N} e(f(n))$. Van der Corput (1921) proved that $S \ll 1/\theta$. Kusmin (1927) sharpened this to $|S| \le 1/\theta$, and Landau (1928) gave the estimate $|S| \le \cot(\pi\theta/2)$. Elementary proofs of all of the these results can be found in a paper of Mordell (1958). The Kusmin-Landau inequality was rediscovered by Herzog and Piranian (1949) in some work on sets of convergence of Taylor series. We have essentially followed them in our proof.

Lemma 2.4, in the special case $H_1 = 0$ and $H_2 = \infty$, is due to van der Corput (1922). Srinivasan (1983) first enunciated the general case.

Despite it simplicity and usefulness, the convexity principle has not always been well appreciated by researchers in exponential sums. For example, it does not appear explicitly in the works of van der Corput, Titchmarsh, or Phillips. For a nice illustration of the power of convexity, see the work of Heath-Brown and Iwaniec (1979) on gaps between primes.

Theorem 2.8 should be compared to Theorem 5.13 of Titchmarsh (1951). The main terms are identical, but the secondary terms in our theorem are smaller. The improvement comes from the use of Lemma 2.4.

The estimate $\zeta(1+it) \ll \log t/\log\log t$ was first proved by Weyl (1921). Stronger estimates can be obtained from Vinogradov's mean-value estimates (see Titchmarsh (1951, Chapter 6) and Ivić (1985, Chapter 6)).

3. THE METHOD OF EXPONENT PAIRS

3.1 INTRODUCTION

In this chapter, we will develop Phillips' method of exponent pairs. We are again striving for upper bounds for sums of the form

$$S = \sum_{n \in I} e(f(n)), \tag{3.1.1}$$

where $I = (a, b]$. However, we now assume that there is some $N > 0$ such that $I \subseteq [N, 2N]$, and our estimates will be in terms of N rather than $|I|$. We will also assume that f is some suitably well-defined function. The exact conditions on f are complicated (see Section 3.3), but the most important feature is that $f'(x)$ is approximately yx^{-s} for some $y > 0$ and some $s > 0$. The bounds for S are expressed in terms of N and $L \overset{\text{def}}{=} yN^{-s}$. (Note that $f' \approx L$.) If $L \leq 1$ then a satisfactory estimate can be obtained from Theorems 2.1 and 2.2, so we concentrate on the case $L \geq 1$. In this case, we seek an estimate of the form

$$S \ll L^k N^l.$$

If we can prove such an estimate, we say that (k, l) is an exponent pair.

Clearly, $(0, 1)$ is an exponent pair. The Phillips' theory consists of two processes - A and B - which produce new exponent pairs from old ones.

The A process is based on Lemma 2.5, the Weyl-van der Corput inequality. We shall see that if (k, l) is an exponent pair then so is

$$A(k, l) = \left(\frac{k}{2k+2}, \frac{k+l+1}{2k+2} \right).$$

The B-process is derived from the Poisson summation formula. From the exponent pair (k, l), the B-process produces the pair

$$B(k, l) = (l - 1/2, k + 1/2).$$

Note that $B(0, 1) = (1/2, 1/2)$; this exponent pair gives the same estimate as Theorem 2.9 in the case $q = 0$. Moreover, $AB(0, 1) = (1/6, 2/3)$, which is Theorem 2.9 in the case $q = 1$. (The product AB is to be interperted as a composition in the usual sense of analysis. Thus $AB(0, 1) = A(1/2, 1/2) = (1/6, 2/3)$.)

The B-process is involutory; i.e. $B^2(k,l) = (k,l)$. Consequently, any exponent pair derivable from the Phillips' theory has the form

$$A^{q_1}BA^{q_2}B\ldots A^{q_k}B(0,1) \tag{3.1.2}$$

or

$$BA^{q_1}BA^{q_2}B\ldots A^{q_k}B(0,1), \tag{3.1.3}$$

where the q_i's are positive integers.

In Chapter 4, we will give several applications of exponent pairs. In these applications, the ultimate result will be expressed as a function of k and l of the form

$$\frac{ak+bl+c}{dk+el+f} \tag{3.1.4}$$

where $a,\ldots,f$ are real numbers. In Chapter 5, we consider the question of how to minimize (3.1.4), or equivalently, how to make an optimal choice of the q_i's in (3.1.2) or (3.1.3).

This chapter is organized in the following way. In Section 3.2, we give several lemmas on exponential integrals that will be used in the proof of the B-process. In Section 3.3, we give heuristic arguments for the A and B processes. Using these heuristics as a guide, we give the formal definition of an exponent pair. Sections 3.4 and 3.5 contain the proofs of the A and B processes respectively.

3.2 LEMMAS ON EXPONENTIAL INTEGRALS

In this section, we prove a number of estimates on exponential integrals. However, the only result that will be used in the other sections of this chapter is Lemma 3.6, which is the main tool necessary to prove the B-process.

LEMMA 3.1. *Assume that f and g are twice differentiable on $[a,b]$. Assume moreover that g/f' is monotonic and that $|f'(x)/g(x)| \geq \lambda$. Then*

$$\int_a^b g(x)e(f(x))dx \ll 1/\lambda.$$

Proof. An integration by parts shows that the left hand side is

$$\frac{1}{2\pi i}\int_a^b \frac{g(x)\,de(f(x))}{f'(x)} = \frac{g(x)e(f(x))}{2\pi i f'(x)}\bigg]_a^b - \frac{1}{2\pi i}\int_a^b e(f(x))\frac{d}{dx}\frac{g(x)}{f'(x)}\,dx$$

$$\ll \frac{1}{\lambda} + \int_a^b \left|\frac{d}{dx}\frac{g(x)}{f'(x)}\right| \,dx.$$

Since g/f' is monotonic, the absolute value bars may be taken outside the integral, and the desired estimate follows.

LEMMA 3.2. *Assume that f has two continuous derivatives and that $|f''(x)| \geq \lambda_2 > 0$ on $[a,b]$. Then*

$$\int_a^b e(f(x))dx \ll \lambda_2^{-1/2}.$$

Proof. Let λ be a number to be chosen later. We write $[a,b] = E_1 \cup E_2$, where

$$E_1 = \{x : |f'(x)| \geq \lambda\} \text{ and } E_2 = \{x : |f'(x)| < \lambda\}.$$

Since f'' has constant sign, E_1 consists of at most two intervals and E_2 consists of at most one interval. By Lemma 3.1,

$$\int_{E_1} e(f(x))dx \ll \frac{1}{\lambda}.$$

Let $E_2 = (c,d)$. Then

$$(d-c)\lambda_2 \leq \left|\int_c^d f''(x)dx\right| = |f'(d) - f'(c)| \leq 2\lambda,$$

so

$$\left|\int_{E_2} e(f(x))dx\right| \leq d - c \ll \lambda/\lambda_2.$$

Combining the above gives

$$\int_a^b e(f(x))dx \ll 1/\lambda + \lambda/\lambda_2;$$

choosing $\lambda = \lambda_2^{1/2}$ completes the proof.

LEMMA 3.3. *Let A and X be positive numbers. Then*

$$\int_{-X}^X e(Ax^2)dx = \frac{e(1/8)}{\sqrt{2A}} + O\left(\frac{1}{AX}\right).$$

Proof. After a couple of obvious changes of variables, we see that it suffices to show that

$$\int_0^X e(x^2)\,dx = \frac{e(1/8)}{2\sqrt{2}} + O\left(\frac{1}{X}\right).$$

Consider the contour integral

$$\int_{\mathbf{C}} e(z^2)dz,$$

where $\mathbf{C}$ is the path which goes along the straight line from 0 to X, along the circular arc from X to $Xe(1/8)$, and along the straight line from $Xe(1/8)$ to 0. Since the integrand is entire, the integral around the contour is 0. Therefore

$$\int_0^X e(x^2)\,dx + \int_{\mathbf{C}_1} e(z^2)dz = e(1/8)\int_0^X e^{-2\pi x^2}\,dx,$$

where $\mathbf{C}_1$ is the circular arc of $\mathbf{C}$. Since $\sin 2\theta \geq 4\theta/\pi$ for $0 \leq \theta \leq \pi/4$, we have

$$\Big|\int_{\mathbf{C}_1}\Big| \leq X\int_0^{\pi/4} \exp(-4\pi X^2 \sin 2\theta)d\theta \leq X\int_0^{\pi/4} \exp(-16X^2\theta)d\theta \ll 1/X.$$

Furthermore,

$$\begin{aligned}\int_0^X e^{-2\pi x^2}\,dx =& \frac{1}{2\sqrt{2\pi}}\int_0^{2\pi X^2} y^{-1/2}e^{-y}dy \\ =& \frac{1}{2\sqrt{2\pi}}\left\{\Gamma(1/2) - \int_{2\pi X^2}^{\infty} y^{-1/2}e^{-y}\,dy\right\} \\ =& \frac{1}{2\sqrt{2}} + O\left(\frac{1}{X}\right).\end{aligned}$$

LEMMA 3.4. *Suppose g is a real valued function with four continuous derivatives on $[a,b]$. Suppose also that $g''(x) \geq \lambda_2 > 0$ and that $g''(x_0) = 0$ for some $x_0 \in [a,b]$. Finally, assume that there are positive constants λ_3 and λ_4 such that*

$$|g^{(3)}(x)| \leq \lambda_3 \text{ and } |g^{(4)}(x)| \leq \lambda_4$$

on $[a,b]$. Then

$$\int_a^b e(g(x))dx = \frac{e(1/8 + g(x_0))}{g''(x_0)^{1/2}} + O(R_1 + R_2), \tag{3.2.1}$$

where

$$R_1 = \min\left(\frac{1}{\lambda_2(x_0 - a)}, \frac{1}{\lambda_2^{1/2}}\right) + \min\left(\frac{1}{\lambda_2(b - x_0)}, \frac{1}{\lambda_2^{1/2}}\right)$$

and

$$R_2 = (b-a)\lambda_4\lambda_2^{-2} + (b-a)\lambda_3^2\lambda_2^{-3}.$$

If the condition $g''(x) \geq \lambda_2 > 0$ is replaced by $g''(x) \leq -\lambda_2 < 0$, then we may replace g with $-g$ and get the same result, save that the main term is now

$$\frac{e(-1/8 + g(x_0))}{|g''(x_0)|^{1/2}}.$$

In our applications of Lemma 3.4, we will be assuming that $\lambda_j \approx FN^{-j}$ for $j = 2, 3$, and 4. In this case, $R_2 = F^{-1}N$.

Proof. We may assume that

$$a + \lambda_2^{-1/2} \leq x_0 \leq b - \lambda_2^{-1/2}, \tag{3.2.2}$$

for otherwise Lemma 3.2 is better that our claimed result.

We multiply both sides of (3.2.1) by $e(-g(x_0))$ to reduce to the case $g(x_0) = 0$. Similarly, we may translate coordinates so that $x_0 = 0$. Let $q(x) = \frac{1}{2}g''(0)x^2$ and $r(x) = g(x) - q(x)$. Since $q(x)$ is the Taylor approximation to g of order 2, we may write

$$r(x) = \frac{1}{2}x^3 \int_0^1 (1-u)^2 g^{(3)}(xu)\, du \tag{3.2.3}$$

Moreover, $r'(x)$ is the Taylor approximation to $g'(x)$ of order 1, so we may write

$$r'(x) = x^2 \int_0^1 (1-u) g^{(3)}(xu)\, du. \tag{3.2.4}$$

We now write

$$\int_a^b e(g(x))dx = \int_a^b e(q(x))dx + \int_a^b e(q(x))\{e(r(x)) - 1\}dx.$$

The first integral can be handled with Lemma 3.3. It then suffices to show that

$$\int_a^b e(q(x))\{e(r(x)) - 1)\}\, dx \ll R_1 + R_2. \tag{3.2.5}$$

Integrating by parts, we see that this integral is

$$\left.\frac{e(q(x))\{e(r(x)) - 1)\}}{2\pi i g''(0)x}\right]_a^b - \\ \frac{1}{g''(0)} \int_a^b e(q(x)) \left\{ \frac{e(r(x))r'(x)}{x} - \frac{e(r(x)) - 1}{2\pi i x^2} \right\} dx. \tag{3.2.6}$$

The first term contributes $\ll |a|^{-1}\lambda_2^{-1} + b^{-1}\lambda_2^{-1} \ll R_1$. We write the integral in (3.2.6) as $T_1 + \frac{1}{2\pi}T_2$, where T_1 arises from the first term in brackets, and T_2 arises from the second term.

Let $h(x) = r'(x)x^{-2}$ and $j(x) = g'(x)x^{-1}$. Then

$$\begin{aligned} T_1 &= \int_a^b \frac{e(g(x))r'(x)}{x}\, dx = \int_a^b \frac{h(x)}{j(x)} e(g(x))g'(x)dx = \int_a^b \frac{h(x)\, de(g(x))}{2\pi i j(x)} \\ &= \left.\frac{h(x)e(g(x))}{2\pi i j(x)}\right]_a^b - \int_a^b \frac{d}{dx}\left(\frac{h(x)}{j(x)}\right) \frac{e(g(x))}{2\pi i}\, dx. \end{aligned}$$

From (3.2.4), we see that

$$h(x) = \int_0^1 (1-u)g^{(3)}(xu)\, du \ll \lambda_3,$$

and

$$h'(x) = \int_0^1 (1-u)ug^{(4)}(xu)\, du \ll \lambda_4.$$

Similarly,

$$j(x) = \int_0^1 g''(xu)\,du \geq \lambda_2 \text{ and } j'(x) = \int_0^1 ug^{(3)}(xu)\,du \ll \lambda_3.$$

Thus

$$\frac{d}{dx}\frac{h}{j} = \frac{h'}{j} - \frac{hj'}{j^2} \ll \frac{\lambda_4}{\lambda_2} + \frac{\lambda_3^2}{\lambda_2^2},$$

and so

$$T_1 \ll \frac{\lambda_3}{\lambda_2} + \frac{\lambda_4(b-a)}{\lambda_2} + \frac{\lambda_3^2(b-a)}{\lambda_2^2}.$$

To treat T_2, we let δ be a parameter to be chosen later, and we set

$$\mathbf{I} = [a,b] \cap [-\delta,\delta]\ ,\ \mathbf{J} = [a,b] - [-\delta,\delta].$$

Note that $\mathbf{I} = [c,d]$, where $c = \max(a,-\delta)$ and $d = \min(b,\delta)$. Define $k(x) = r(x)x^{-3}$. From (3.2.3), we see that $k(x) \ll \lambda_3$ and $k'(x) \ll \lambda_4$. Set $m(x) = (e(x)-1)/x$; then $m(x) \ll 1$ and $m'(x) \ll 1$. The contribution of the interval $\mathbf{I}$ to T_2 is

$$\begin{aligned}
&\int_c^d e(q(x))xm(r(x))k(x)dx \\
&= \left.\frac{e(q(x))m(r(x))k(x)}{2\pi i g''(0)}\right]_c^d \\
&\quad - \frac{1}{2\pi i g''(0)}\int_c^d e(q(x))\{m'(r(x))r'(x)k(x) + m(r(x))k'(x)\}dx \\
&\ll \lambda_2^{-1}\lambda_3 + \lambda_2^{-1}\lambda_3^2\delta^3 + \lambda_2^{-1}\lambda_4(b-a).
\end{aligned}$$

For the interval $\mathbf{J}$, we apply Lemma 3.1 to the integrals

$$\int_{\mathbf{J}} \frac{e(g(x))}{x^2}\,dx \text{ and } \int_{\mathbf{J}} \frac{e(q(x))}{x^2}\,dx.$$

In the first integral, we are lead to consider the function $g'(x)x^2$. The absolute value of this function is $\geq \lambda_2\delta^3$, and the expression is monotonic since $(g'(x)x^2)' = g''(x)x^2 + 2xg'(x) > 0$. By Lemma 3.1,

$$\int_{\mathbf{J}} \frac{e(g(x))}{x^2}\,dx \ll \lambda_2^{-1}\delta^{-3}.$$

Similarly, since $q'(x)x^2 = g''(0)x^3$ is monotonic,

$$\int_{\mathbf{J}} \frac{e(q(x))}{x^2}\,dx \ll \lambda_2^{-1}\delta^{-3}.$$

We take $\delta = \lambda_3^{-1/3}$ and combine this with our estimate of T_1 to see that the integral in (3.2.5) is

$$\ll R_1 + \lambda_2^{-1}(T_1 + T_2) \ll R_1 + \lambda_2^{-2}\lambda_3 + \lambda_2^{-2}\lambda_4(b-a) + \lambda_2^{-3}\lambda_3^2(b-a). \qquad (3.2.7)$$

Put $U = \lambda_2^{-1}(b-a)^{-1}$. From (3.2.2), we see that $U \ll R_1$. Since

$$\lambda_2^{-2}\lambda_3 = U^{1/2}(\lambda_2^{-3}\lambda_3^2(b-a))^{1/2} \ll \max(R_1, \lambda_2^{-3}\lambda_3^2(b-a)),$$

the second term in (3.2.7) may be omitted. This completes the proof.

For the next lemma, recall that we defined $[x]$ to be the greatest integer $\leq x$, and $\psi(x) = x - [x] - 1/2$.

LEMMA 3.5. *Suppose f has two continuous derivatives on $[a,b]$. Suppose also that f' is decreasing, that H_1 and H_2 are integers such that $H_1 < f'(x) < H_2$, and that $H = H_2 - H_1 \geq 2$. Then*

$$\sum_{n\in I} e(f(n)) = \sum_{H_1\leq h\leq H_2} \int_a^b e(f(x) - hx)\,dx + O(\log H).$$

Proof. The left-hand side is unchanged if we replace $f(n)$ by $f(n) - kn$ for any integer k. Therefore, we may normalize to the case $H_1 < 0 < H_2$.

Using Riemann-Stieltjes integration, we get

$$\sum_{n\in I} e(f(n)) = \int_a^b e(f(x))dx - \int_a^b e(f(x))d\psi(x) + O(1). \qquad (3.2.8)$$

Integration by parts shows that

$$-\int_a^b e(f(x))d\psi(x) = 2\pi i \int_a^b e(f(x))f'(x)\psi(x)dx + O(1).$$

The Fourier expansion

$$\psi(x) = \sum_{h\neq 0} \frac{e(-hx)}{2\pi i h}$$

is boundedly convergent for all x, so

$$2\pi i \int_a^b e(f(x))f'(x)\psi(x)dx = \sum_{h\neq 0} \frac{1}{h} \int_a^b e(f(x) - hx)f'(x)\,dx. \qquad (3.2.9)$$

If $h < H_1$ or $h > H_2$ then $f'(x) - h$ is monotonic and non-zero on $[a,b]$. Moreover, $f'(x)/(f'(x) - h)$ is monotonic, so

$$\int_a^b e(f(x) - hx)f'(x)dx \ll \frac{|f'(b)|}{|f'(b) - h|} + \frac{|f'(a)|}{|f'(a) - h|}.$$

by 3.1. When $h > H_2$, the above is $\ll H_2/(h - H_2)$. Thus these terms contribute

$$\ll \sum_{h>H_2, h\neq 0} \frac{H_2}{h(h-H_2)} \ll \sum_{h>H_2, h\neq 0} \left\{ \frac{1}{h-H_2} - \frac{1}{h} \right\} \ll \log H$$

to (3.2.8). The same upper bound holds for the terms with $h < H_1$. We integrate the remaining terms by parts to get

$$\begin{aligned}
&\sum_{\substack{H_1 \le h \le H_2 \\ h\neq 0}} \frac{1}{h} \int_a^b f'(x) e(f(x) - hx) dx \\
&= \sum_{\substack{H_1 \le h \le H_2 \\ h\neq 0}} \frac{1}{2\pi i h} \int_a^b e(-hx) de(f(x)) \\
&= \sum_{\substack{H_1 \le h \le H_2 \\ h\neq 0}} \frac{e(f(x) - hx)}{2\pi i h} \Bigg]_a^b + \sum_{\substack{H_1 \le h \le H_2 \\ h\neq 0}} \int_a^b e(f(x) - hx) dx \\
&= \sum_{\substack{H_1 \le h \le H_2 \\ h\neq 0}} \int_a^b e(f(x) - hx) dx + O(\log H).
\end{aligned}$$

Combining the above with (3.2.8) and (3.2.9) yields

$$\sum_{n\in I} e(f(n)) = \int_a^b e(f(x)) dx + \sum_{\substack{H_1 \le h \le H_2 \\ h\neq 0}} \int_a^b e(f(x) - hx)\, dx + O(\log H).$$

Since $H_1 < 0 < H_2$, this completes the proof.

LEMMA 3.6. *Suppose that f has four continuous derivatives on $[a, b]$, and that $f'' < 0$ on this interval. Suppose further that $[a, b] \subseteq [N, 2N]$ and that $\alpha = f'(b)$ and $\beta = f'(a)$. Assume that there is some $F > 0$ such that*

$$f^{(2)}(x) \approx FN^{-2}\,,\ f^{(3)}(x) \ll FN^{-3}\,,\ \text{ and } f^{(4)}(x) \ll FN^{-4}$$

for x in $[a, b]$. Let x_ν be defined by the relation $f'(x_\nu) = \nu$, and let $\phi(\nu) = -f(x_\nu) + \nu x_\nu$. Then

$$\sum_{n\in I} e(f(n)) = \sum_{\alpha \le \nu \le \beta} \frac{e(-\phi(\nu) - 1/8)}{|f''(x_\nu)|^{1/2}} + O(\log(FN^{-1} + 2) + F^{-1/2} N).$$

Proof. We may assume that $F \ge 1$, for the theorem is trivial otherwise. Let $H_1 = [\alpha] - 1$ and $H_2 = [\beta] + 1$. From the previous two lemmas, we see that

$$\sum_{n\in I} e(f(n)) = \sum_{H_1 \le \nu \le H_2} \frac{e(-\phi(\nu) - 1/8)}{|f''(x_\nu)|^{1/2}} + O(E), \tag{3.2.10}$$

where

$$E = \sum_{H_1 \le \nu \le H_2} \{F^{-1}N + \min(|f'(a) - \nu|^{-1}, F^{-1/2}N) + \min(|f'(b) - \nu|^{-1}, F^{-1/2}N)\}.$$

Note that

$$H_2 - H_1 \ll \left|\int_a^b f''(x)dx\right| + 1 \ll FN^{-1} + 1,$$

so

$$\sum_{H_1 \le \nu \le H_2} F^{-1}N \ll F^{-1}N(FN^{-1} + 1) \ll 1 + F^{-1}N.$$

The second term is majorized by $F^{-1/2}N$ since we are assuming that $F \ge 1$. Note also that for $\nu < H_2 - 1$,

$$|f'(a) - \nu| \ge H_2 - 1 - \nu$$

so

$$\sum_{H_1 \le \nu \le H_2} \min(|f'(a) - \nu|^{-1}, F^{-1/2}N) \ll F^{-1/2}N + \sum_{H_1 \le \nu < H_2 - 1} (H_2 - 1 - \nu)^{-1}$$
$$\ll F^{-1/2}N + \log(FN^{-1} + 2).$$

The same upper bound holds for the sum

$$\sum_{H_1 \le \nu \le H_2} \min(|f'(b) - \nu|^{-1}, F^{-1/2}N).$$

Finally, if we change the range of summation in (3.2.6) to $\alpha < \nu \le \beta$, this changes the sum by $\ll F^{-1/2}N$, and this may be absorbed into the error term.

3.3 HEURISTIC ARGUMENTS AND DEFINITIONS

We begin this section by giving some heuristic arguments for the A and B processes. Then, partly guided by these heuristics, we give the formal definition of exponent pairs.

Let S be as defined in (3.1.1). Recall the Weyl-van der Corput inequality, which states that if $H \le N$ then

$$|S|^2 \ll \frac{N^2}{H} + \frac{N}{H} \sum_{1 \le h \le H} |S_1(h)|$$

where

$$S_1(h) = \sum_{\substack{n \\ a < n \le b\ ,\ a < n+h \le b}} e(f_1(n;h)), \tag{3.3.1}$$

and $f_1(n;h) = f(n+h) - f(n)$. If $f' \approx L$, then $f_1' \approx L|h|N^{-1}$. If we assume that the exponent pair (k,l) can be applied to $S_1(h)$, then we obtain

$$|S|^2 \ll N^2H^{-1} + NH^{-1}\sum_{1\le h\le H}(L|h|N^{-1})^kN^l \ll N^2H^{-1} + H^kL^kN^{1-k+l}.$$

Upon choosing H to make the two terms equal, we get

$$S \ll L^{k/(2k+2)}N^{(k+l+1)/(2k+2)}.$$

For the B-process, we use Lemma 3.6 to get

$$S = e(-1/8)\sum_{\alpha\le\nu\le\beta}\frac{e(-\phi(\nu))}{|f''(x_\nu)|^{1/2}} + \text{error terms.} \tag{3.3.2}$$

For the purposes of this heuristic argument, we will ignore the error terms. Now $\phi'(\nu) = x_\nu \approx N$, and $f''(x_\nu) \approx LN^{-1}$. By partial summation, we see that

$$\sum_{\alpha\le\nu\le\beta}\frac{e(\phi(\nu))}{|f''(x_\nu)|^{1/2}} \ll L^{-1/2}N^{1/2}\max_{\beta'\le\beta}|\sum_{\alpha<\nu\le\beta}e(\phi(\nu))|$$

If we assume that the exponent pair (k,l) may be applied to the last sum, we get

$$S \ll L^{-1/2}N^{1/2}N^kL^l = L^{l-1/2}N^{k+1/2}.$$

To make these arguments rigorous, we need to work with a suitable class of functions $\mathbf{F}$. This class of functions $\mathbf{F}$ should have the property that if $f \in \mathbf{F}$ then the auxiliary functions f_1 and ϕ (of (3.3.1) and (3.3.2) respectively) are also in $\mathbf{F}$.

Another requirement for our class of functions $\mathbf{F}$ is that it contains those functions f which satisfy $f'(x) = yx^{-s}$ for some $s > 0$. For these functions we have

$$f^{(p+1)}(x) = (-1)^p(s)_pyx^{-s-p}.$$

Here, $(s)_p$ is defined recursively by the relations

$$(s)_0 = 1, (s)_{k+1} = (s+k)(s)_k.$$

Note that when $p \ge 1$, $(s)_p = s(s+1)\ldots(s+p-1)$.

Definition. Let N, y, s, and ϵ be positive real numbers with $\epsilon < 1/2$, and let P be a non-negative integer. Define $\mathbf{F}(N,P,s,y,\epsilon)$ to be the set of functions f such that

(1) f is defined and has P continuous derivatives on some interval $[a,b]$, with $[a,b] \subseteq [N,2N]$,

(2) if $0 \le p \le P-1$ and $a \le x \le b$ then

$$|f^{(p+1)}(x) - (-1)^p (s)_p y x^{-s-p}| < \epsilon (s)_p y x^{-s-p}. \tag{3.3.3}$$

Definition. Let k and l be real numbers such that $0 \le k \le 1/2 \le l \le 1$. Suppose that for every $s > 0$, there is some $P = P(k,l,s)$ and some $\epsilon = \epsilon(k,l,s) < 1/2$ such that for every $N > 0$, every $y > 0$, and every $f \in \mathbf{F}(N, P, s, y, \epsilon)$, the estimate

$$\sum_{n \in I} e(f((n)) \ll (yN^{-s})^k N^l + y^{-1} N^s, \tag{3.3.4}$$

holds. Here it is also assumed that f is defined on $[a,b]$ and the implied constant depends only on k, l, and s. We then say that (k,l) is an exponent pair.

In proving that (k,l) is an exponent pair, we may always assume that $yN^{-s} \ge 1$. For if $yN^{-s} < 1/2$, then Theorem 2.1 yields

$$\sum_{n \in I} e(f((n)) \ll y^{-1} N^{-s}.$$

If $1/2 \le yN^{-s} < 1$ then Theorem 2.2 yields

$$\sum_{n \in I} e(f((n)) \ll N^{1/2},$$

and $N^{1/2} \le (yN^{-s})^k N^l$ since $l \ge 1/2$.

We close this section with an explanation of why the conditions $0 \le k \le 1/2 \le l \le 1$ are included in the definition of exponent pairs. The condition $l \ge 1/2$ is no essential restriction since no pair (k,l) which satisfies (3.3.4) can have $l < 1/2$. To see why, consider the sum

$$S(t) = \sum_{N < t \le 2N} e\left(\frac{t}{n}\right).$$

By Theorem 2 of Appendix A,

$$\int_T^{2T} |S(t)|^2 dt \ge (T - 4N^2) N.$$

It follows that if $T \ge 8N^2$ then there is some t, $T \le t \le 2T$, such that $|S(t)| \gg N^{1/2}$.

We also claim that there is a sequence $\{t_\nu\}_{\nu=1}^{\infty}$ such that $t_\nu \to \infty$ and $|S(t_\nu)| = N$. Take $t_\nu = \nu$l.c.m. $\{1, 2, \ldots, 2N\}$. Then $t_\nu / n \in \mathbf{Z}$ whenever $N < n \le 2N$, and so $S(t_\nu) = N$. This shows that no exponent pair (k,l) can have $k < 0$, and any exponent pair that has $k = 0$ must have $l = 1$.

There is no need to ever take $l > 1$ in an exponent pair, for an exponent pair with $k \ge 0$ and $l > 1$ would yield an estimate worse than the trivial estimate provided by the exponent pair $(0,1)$. Similarly, there is no need to take $k > 1/2$, for then (k,l) would provide a weaker estimate than the pair $(1/2,1/2)$ of Theorem 2.2.

Finally, we observe that if $(k,1/2)$ is an exponent pair, then $k = 1/2$. Assume that $(k,1/2)$ is an exponent pair. Let H be a positive integer. Let t be defined by the equation $t^2 = \text{l.c.m.}\{1,2,\ldots,H\}$, and let $N = t^2H^{-2}$. Define $f(x) = 2tx^{-2}$. Then

$$\sum_{N<n\le 2N} e(f(n)) \ll H^k N^{1/2} + H^{-1}. \tag{3.3.5}$$

On the other hand, we see from Lemma 3.6 that

$$\sum_{N<n\le 2N} e(f(n)) = \sum_{\frac{1}{\sqrt{2}}tN^{-1/2}<\nu\le tN^{-1/2}} \frac{e(-t^2/\nu - 1/8)}{|f''(x_\nu)|^{1/2}} + O(\log(2+H^{-1})) + O(H^{-1}).$$

Note that $tN^{-1/2} = H$ and that $f''(x_\nu) \approx H/N$. Our definition of t insures that $t^2/\nu \in \mathbf{Z}$ for all ν in the range of summation; therefore,

$$\sum_{N<n\le 2N} e(f(n)) \gg H^{1/2}N^{1/2}. \tag{3.3.6}$$

Estimates (3.3.5) and (3.3.6) are inconsistent unless $k \ge 1/2$, so the result is proved.

3.4 PROOF OF THE A-PROCESS

Our first lemma shows that the auxiliary function f_1 of (3.1.1) is in an appropriate $\mathbf{F}$ set.

LEMMA 3.7. *Suppose $f \in \mathbf{F}(N,P,s,y,\epsilon)$, that f is defined on $[a,b]$, and that $1 \le h < \min(b-a, 2\epsilon N/(s+P))$. Define*

$$f_1(x) = f(x) - f(x+h).$$

Then f_1 is defined on $[a,b-h]$ and $f_1 \in \mathbf{F}(N,P-1,s+1,shy,3\epsilon)$.

Proof. Let

$$F(x) = \begin{cases} yx^{1-s}(1-s)^{-1} & \text{if } s \ne 1; \\ y\log x & \text{if } s = 1. \end{cases}$$

Condition (3.3.2) may be written as

$$|f^{(p+1)}(x) - F^{(p+1)}(x)| < \epsilon|F^{(p+1)}(x)|.$$

Define $F_1(x) = F(x) - F(x+h)$ and $G(x) = -hyx^{-s}$. We will show that

$$|f_1^{(p+1)}(x) - F_1^{(p+1)}(x)| < \epsilon|F_1^{(p+1)}(x)| \tag{3.4.1}$$

and that

$$|F_1^{(p+1)}(x) - G^{(p+1)}(x)| < \epsilon |G^{(p+1)}(x)| \tag{3.4.2}$$

when $0 \le p \le P-2$. The desired result will then follow readily.

From our definitions of f_1 and F_1, we see that

$$|f_1^{(p+1)}(x) - F_1^{(p+1)}(x)| = \left| \int_x^{x+h} f^{(p+2)}(u) - F^{(p+2)}(u)\, du \right|.$$

When $a \le x \le b-h$ and $0 \le p \le P-2$, the above is bounded by

$$\epsilon \int_x^{x+h} |F^{(p+2)}(u)| du = \epsilon |F_1^{(p+1)}(x)|.$$

This proves (3.4.1). For (3.4.2), we observe that the left-hand side is equal to

$$y(s)_{p+1} \left| \int_x^{x+h} (u^{-s-p-1} - x^{-s-p-1}) du \right| = y(s)_{p+2} \left| \int_x^{x+h} \int_x^u w^{-s-p-2}\, dw\, du \right|,$$

and this in turn is

$$\le \frac{1}{2}(s)_{p+2} y h^2 x^{-s-p-2} < \epsilon (s+1)_p s h y x^{-s-p-1}$$

provided $h < 2\epsilon x/(s+p+1)$. This last condition is guaranteed by our hypothesis $h < 2\epsilon N/(s+P)$.

To complete the proof, we note that

$$|F_1^{(p+1)}(x)| < (1+\epsilon)|G^{(p+1)}(x)|$$

follows from (3.4.2). Combining this with (3.4.1) and (3.4.2) yields

$$|f_1^{(p+1)}(x) - G^{(p+1)}(x)| < (2\epsilon + \epsilon^2)|G^{(p+1)}(x)|.$$

Note that $2\epsilon + \epsilon^2 < 3\epsilon$ since $\epsilon < 1$.

THEOREM 3.8. *If (k, l) is an exponent pair, then*

$$(\kappa, \lambda) = A(k,l) = (k/(2k+2), (k+l+1)/(2k+2))$$

is an exponent pair.

Proof. First, we observe that since $0 \le k$ and $1/2 \le l \le 1$, we have

$$0 \le \kappa = \frac{k}{2k+2} < \frac{k+1}{2k+2} = \frac{1}{2} \text{ and } \frac{1}{2} \le \lambda = \frac{1}{2} + \frac{l}{2k+2} \le 1.$$

Now let y, N, and s be positive. We need to show that there exists $P_1 > 0$ and $\epsilon_1 > 0$ such that if $f \in \mathbf{F}(N, P_1, s, y, \epsilon_1)$ and $L = yN^{-s} \geq 1$ then

$$\sum_{n \in I} e(f(n)) \ll L^{\kappa} N^{\lambda}.$$

Our proof breaks into two cases: (i) $L \geq \log N$, (ii) $1 \leq L < \log N$. We begin with case (i).

Since (k, l) is an exponent pair, we know that there exists $P > 0$ and $\epsilon (0 < \epsilon < 1/2)$ such that if $f \in \mathbf{F}(N, P, s, y, \epsilon)$ then

$$\sum_{n \in I} e(f(n)) \ll (yN^{-s})^k N^l + y^{-1} N^s.$$

We will show that we may take $P_1 = P + 1$ and $\epsilon_1 = \epsilon/3$.

Assume that $H \leq \min(b - a, 2\epsilon N/(s + P))$. In the notation of (2.3.4), we have

$$|S|^2 \ll \frac{N^2}{H} + \frac{N}{H} \sum_{1 \leq h \leq H} |S_1(h)|.$$

By Lemma 3.7, $f_1 \in \mathbf{F}(N, P, shy, s + 1, \epsilon)$, so the exponent pair (k, l) may be applied to $S_1(h)$. Consequently,

$$\begin{aligned} |S|^2 \ll & H^{-1} N^2 + H^{-1} N \sum_{1 \leq h \leq H} \{(hLN^{-1})^k N^l + h^{-1} L^{-1} N\} \\ \ll & H^{-1} N^2 + H^k L^k N^{l-k+1} + H^{-1} L^{-1} N^2 \log N. \end{aligned}$$

Since we are assuming that $L \geq \log N$, the first term dominates the third. Applying Lemma 2.4 and using the upper bound $H \leq \min(b - a, 2\epsilon N/(s + P))$ yields

$$S \ll L^{\kappa} N^{\lambda} + N(b - a)^{-1/2}.$$

If the first term dominates, we are done. Otherwise, we employ the trivial estimate to get

$$S \ll \min(N(b - a)^{-1/2}, b - a) \ll N^{2/3}.$$

Since $k \leq 1/2$ and $l \geq 1/2$, we have

$$\lambda = \frac{1}{2} + \frac{l}{2k + 2} \geq \frac{2}{3},$$

and the desired estimate $S \ll N^{\kappa} L^{\lambda}$ follows.

The remaining case where $1 \leq L \leq \log N$ is easily dispatched. By Lemma 2.2,

$$S \ll N^{1/2} (\log N)^{1/2} \ll N^{2/3} \ll L^{\kappa} N^{\lambda}.$$

3.5 PROOF OF THE B-PROCESS

As in the previous section, we begin by showing that the needed auxiliary function is in an appropriate **F**-set.

LEMMA 3.9. *Suppose $f \in \mathbf{F}(N,P,s,y,\epsilon)$, and f is defined on the interval $[a,b]$. Let $\alpha = f'(b)$ and $\beta = f'(a)$. For $\nu \in [\alpha,\beta]$, let x_ν be defined by the relation $f'(x_\nu) = \nu$, and define $\phi(\nu) = \nu x_\nu - f(x_\nu)$. Let $\sigma = 1/s$ and $\eta = y^\sigma$. Then there is some constant $C = C(s,P)$ such that*

$$|\phi^{p+1}(\nu) - (-1)^p(\sigma)_p\eta\nu^{-\sigma-p}| < C\epsilon(\sigma)_p\eta\nu^{-\sigma-p}$$

whenever $0 \le p \le P-1$ and $\alpha \le \nu \le \beta$.

Proof. As in the proof of Lemma 3.7, let

$$F(x) = \begin{cases} yx^{1-s}(1-s)^{-1} & \text{if } s \neq 1; \\ y\log x & \text{if } s = 1. \end{cases}$$

Condition (3.3.1) may be written as

$$|f^{p+1}(x) - F_{p+1}(x)| < \epsilon|F_{p+1}(x)|. \tag{3.5.1}$$

Now define X_ν by the relation $F'(X_\nu) = \nu$, and define $\Phi(\nu) = \nu X_\nu - F(X_\nu)$. It is easy to see that $x_\nu = \phi'(\nu)$ and $X_\nu = \Phi'(\nu) = \eta\nu^{-\sigma}$. To complete the proof, we will show that there is some constant $C = C(s,P)$ such that

$$|\phi^{p+1}(\nu) - \Phi_{p+1}(\nu))| < C\epsilon\Phi_{p+1}(\nu)) \tag{3.5.2}$$

whenever $0 \le p \le P-1$ and $\alpha \le \nu \le \beta$.

From our definitions, we note that $f'(\phi'(\nu)) = \nu$. By differentiating both sides with respect to ν, we get

$$\phi''(\nu) = \frac{1}{f''(x_\nu)}.$$

An easy induction argument shows that for $p \ge 1$, we can find constants $w(u_1,\ldots,u_{p-1})$ such that

$$\phi^{p+1}(\nu) = \frac{1}{(f''(x_\nu))^{2p-1}}\sum w(u_1,\ldots,u_{p-1})f^{(u_1+1)}(x_\nu)\ldots f^{(u_{p-1}+1)}(x_\nu), \tag{3.5.3}$$

where the sum is over all $u_1,\ldots,u_{p-1}$ such that $1 \le u_j \le p$ and $u_1+\ldots+u_{p-1} = 2p-2$. The constants $w(u_1,\ldots,u_{p-1})$ depend only on $u_1,\ldots,u_{p-1}$ and not on f, so we also have

$$\Phi_{p+1}(\nu)) = \frac{1}{(F''(x_\nu))^{2p-1}}\sum w(u_1,\ldots,u_{p-1})F^{(u_1+1)}(x_\nu)\ldots F^{(u_{p-1}+1)}(x_\nu). \tag{3.5.4}$$

From (3.5.1), we see that

$$|f'(x_\nu) - F'(x_\nu)| = |\nu - yx_\nu^{-s}| < \epsilon y x_\nu^{-s}.$$

Consequently,

$$(1-\epsilon)^\sigma \eta \nu^{-\sigma} < x_\nu < (1+\epsilon)^\sigma \eta \nu^{-\sigma},$$

and so

$$|x_\nu - X_\nu| \ll \epsilon \eta v^{-\sigma}.$$

(Here, and in the remainder of this proof, all $<<$ —constants depend at most on s and P.) From (3.5.1), we see that if $0 < p < P-1$ then

$$|f^{p+1}(x_\nu) - F_{p+1}(x_\nu)| < \epsilon |F_{p+1}(x_\nu)| \ll \epsilon y N^{-s-p}$$

Furthermore, there is some t in the open interval with endpoints x_ν and X_ν such that

$$|F_{p+1}(x_\nu) - F_{p+1}(X_\nu)| = |F^{(p+2)}(t)(X_\nu - x_\nu)| \ll \epsilon y N^{-s-p-1}\eta\nu^{-\sigma} \ll \epsilon y N^{-s-p}.$$

The last two estimates together yield

$$|f^{p+1}(x_\nu) - F_{p+1}(X_\nu)| \ll \epsilon y N^{-s-p}.$$

Upon combining this with (3.5.3) and (3.5.4), we find that

$$|\phi^{p+1}(\nu) - \Phi_{p+1}(\nu))| \ll \frac{\epsilon}{|F''(X_\nu)|^{2p-1}} \sum |w(u_1, \ldots, u_{p-1})| y^{p-1} X_\nu^{-s(p-1)-(u_1+\ldots+u_{p-1})}$$

$$\ll \frac{\epsilon}{|yX_\nu^{-s-1}|^{2p-1}} \ll y^{p-1} X_\nu^{-s(p-1)-(u_1+\ldots+u_{p-1})} \ll \epsilon y^{-p} N^{ps+1} \ll \epsilon |\Phi_{p+1}(\nu))|.$$

This proves (3.5.2), and thus completes the proof of the lemma.

In terms of our definition of $\mathbf{F}$, Lemma 3.9 says that if $f \in \mathbf{F}(N, P, s, y, \epsilon)$ then the restriction of ϕ to the interval $[\alpha, \beta] \cap [J, 2J]$ is in $\mathbf{F}(J, P, \sigma, \eta, C\epsilon)$ for any J, $\alpha \le J \le \beta$.

THEOREM 3.10. *If (k, l) is an exponent pair then*

$$(\kappa, \lambda) = B(k, l) = (l - 1/2, k + 1/2)$$

is an exponent pair.

Proof. First, we observe that $0 \le \kappa \le 1/2 \le \lambda \le l \le 1$ follows immediately from $0 \le k \le 1/2 \le l \le 1$. Moreover, if $l = 1/2$, then, by the remarks at the end of Section 3.3, we have $k = 1/2$. It follows that $(\kappa, \lambda) = (0, 1)$ is an exponent pair by the trivial estimate. We may henceforth assume that $l \ge 1/2$ and $\kappa > 0$.

Assume that $y > 0, s > 0, N > 0$, and that $L = yN^{-s} \geq 1$. We want to find P_1 and ϵ_1 such that if $f \in \mathbf{F}(N, P_1, s, y, \epsilon_1)$ then

$$S = \sum_{n \in I} e(f(n)) \ll L^{\kappa} N^{\lambda}.$$

Since (k, l) is an exponent pair, we know that there exist $P > 0$ and $\epsilon (0 < \epsilon < 1/2)$ such that if $f \in \mathbf{F}(N, P, s, y, \epsilon)$ then (3.3.4) holds. We will show that we may take $P_1 = P$ and $\epsilon_1 = \epsilon / C$, where $C = C(s, P)$ is the constant occurring in Lemma 3.9

Since f satisfies the hypothesis of Lemma 3.6 with $F = LN$, we may write

$$S = \sum_{\alpha \leq \nu \leq \beta} \frac{e(-\phi(\nu) - 1/8)}{|f''(x_\nu)|^{1/2}} + O(\log(2L) + L^{-1/2} N^{1/2}). \tag{3.5.5}$$

By the previous lemma,

$$T(w) \stackrel{\text{def}}{=} \sum_{\alpha < \nu \leq w} e(\phi(\nu)) \ll (\eta J^{-\sigma})^k J^l + \eta^{-1} J^{\sigma} \ll N^k J^l + N^{-1}.$$

Consequently, the sum in (3.5.5) is

$$\begin{aligned} \int_{\alpha}^{\beta} |f''(x_w)|^{-1/2} d\overline{T(w)} &= \overline{T(w)} |f''(x_w)|^{-1/2} \Big]_{\alpha}^{\beta} - \int_{\alpha}^{\beta} \overline{T(w)} \frac{d}{dw} |f''(x_w)|^{-1/2}\, dw \\ &\ll (N^k L^l + N^{-1}) \{ (LN^{-1})^{-1/2} + \int_{\alpha}^{\beta} |\frac{d}{dw} |f''(x_w)|^{-1/2}|\, dw \} \\ &\ll L^{\kappa} N^{\lambda} + L^{-1/2} N^{-1/2}. \end{aligned}$$

All together, we get

$$S \ll L^{\kappa} N^{\lambda} + \log(2L) + L^{-1/2} N^{1/2}.$$

Since we are assuming that $\kappa > 0$ and $L \geq 1$, the first term dominates, and the result is proved.

3.6 NOTES

Here, we have presented the method of exponent pairs along very much the same lines as Phillips (1934). One difference is in the use of Lemma 3.4, which is somewhat stronger than the corresponding result of Phillips. Several variants of this result have appeared in the literature; see, for example Lemma 4.6 and pages 90-91 of Titchmarsh (1951), Lemma 10 of Heath-Brown (1983), and pages 86-91 of Vinogradov (1985). The proof given here is due to H.L. Montgomery (unpublished).

4. APPLICATIONS OF EXPONENT PAIRS

4.1 THE RIEMANN ZETA-FUNCTION

For our first application of exponent pairs, we shall improve the results we obtained in Chapter 2 on the Riemann zeta-function.

THEOREM 4.1. *Suppose that $t \geq 3$, and that (k,l) is an exponent pair such that $k + 2l \geq 3/2$. Let $\theta(k,l) = (2k + 2l - 1)/4$. Then*

$$\zeta(1/2 + it) \ll t^{\theta(k,l)} \log t.$$

In particular,

$$\zeta(1/2 + it) \ll t^{27/164} \log t.$$

Proof. From Lemmas 2.10 and 2.11, we see that it suffices to show that

$$N^{-1/2} \sum_{N<n\leq N_1} n^{-it} \ll t^{\theta(k,l)},$$

whenever $1 \leq N \leq t$ and $N_1 \leq 2N$. Using the exponent pairs (k,l) and $B(k,l) = (l - 1/2, k + 1/2)$, we get

$$N^{-1/2} \sum_{N<n\leq N_1} n^{-it} \ll \min\{t^k N^{l-k-1/2}, t^{l-1/2} N^{k-l+1/2}\}.$$

This together with convexity gives the desired result. The exponent 27/164 follows by taking $(k,l) = ABAB^3(0,1) = (\frac{11}{82}, \frac{57}{82})$.

We remark that 27/164 is not the smallest value for $\theta(k,l)$ that can be obtained from the A and B processes. In Chapter 5, we will give an algorithm for finding the smallest such value.

Our next result sharpens Theorem 2.12.

THEOREM 4.2. *Let $q \geq 1$ be an integer. Define $Q = 2^q$, $\sigma_q = 1 - (q + 2)/(4Q - 2)$, and*

$$\eta_q = \frac{1}{4Q - 2} \cdot \frac{240Qq + 224Q + 128}{240Qq + 225Q + 128}.$$

If $t \geq 3$ then

$$\zeta(\sigma_q + it) \ll t^{\eta_q} \log t.$$

Proof. We see from Lemmas 2.10 and 2.11 that it suffices to show that

$$S = N^{-\sigma_q} \sum_{N<n\leq N_1} n^{-it} \ll t^{\eta_q},$$

whenever $1 \leq N \leq t$ and $N_1 \leq 2N$.

Let

$$(k_q, l_q) = A^q BABA^4B(0,1) = \left(\frac{16}{120Q-32}, \frac{120Q-16q-63}{120Q-32}\right),$$

and let $x_q = k_q - l_q$. It is easy to see that $x_q - 1 < \sigma_q < x_q$. Moreover,

$$S \ll \min\{t^{k_q} N^{x_q-\sigma_q}, t^{k_q-1} N^{x_q-1-\sigma_{q-1}}\},$$

and the estimate $S \ll t^{\eta_q}$ follows by convexity.

4.2 SUMS INVOLVING ψ

In most of the applications of this chapter, we encounter sums of the form

$$\sum_{n\in I} \psi(f(n)),$$

where f is a function satisfying the conditions given in the definition of exponent pairs and $\psi(x) = x - [x] - 1/2$. The following lemma gives a convenient estimate for such sums.

LEMMA 4.3. *Suppose (k,l) is an exponent pair, and that $P = P(k,l)$ and $\epsilon = \epsilon(k,l)$ are the corresponding parameters guaranteed by the definition of exponent pairs. If $f \in \mathbf{F}(N,P,s,y,\epsilon)$ and f is defined on $[a,b]$, then*

$$\sum_{n\in I} \psi(f(n)) \ll y^{k/(k+1)} N^{((1-s)k+l)/(k+1)} + y^{-1}N^s.$$

Proof. Let $J \geq 1$ be a parameter to be chosen later. We use Vaaler's trigonometric polynomial approximation to to ψ, which is given in Appendix A as Theorem A.6. For our purposes here, it is convenient to interpert Vaaler's theorem as stating that there exist coefficients $\gamma(j)$ such that

$$\psi(w) \leq J^{-1} + \sum_{1\leq|j|\leq J} \gamma(j)e(jw)$$

and $\gamma(j) \ll |j^{-1}|$. We therefore obtain

$$\sum_{n\in I} \psi(f(n)) \ll NJ^{-1} + \sum_{1\leq j\leq J} j^{-1} \Big| \sum_{a<n\leq b} e(jf(n)) \Big|.$$

Applying the exponent pair (k,l) to the inner sum yields

$$\sum_{n\in I} \psi(f(n)) \ll NJ^{-1} + \sum_{1\leq j\leq J} j^{-1}\{(jyN^{-s})^k N^l + j^{-1}y^{-1}N^s\}$$
$$\ll NJ^{-1} + J^k N^{l-ks} y^k + y^{-1}N^s.$$

Choosing $J^{k+1} = y^{-k}N^{sk-l+1}$ gives the result.

4.3 THE DIRICHLET DIVISOR PROBLEM

Let $d(n)$ denote the number of divisors of n. Dirichlet used an elementary argument to show that

$$\sum_{n\leq x} d(n) = x(\log x + 2\gamma - 1) + O(x^{1/2}), \tag{4.3.1}$$

where γ is Euler's constant; i.e.

$$\gamma = \lim_{n\to\infty} \Big(\sum_{n\leq N} \frac{1}{n} - \log N\Big). \tag{4.3.2}$$

Dirichlet's result naturally leads to the question of how small the error term in (4.3.1) can be made. The usual convention is to define

$$\Delta(x) = \sum_{n\leq x} d(n) - x(\log x + 2\gamma - 1). \tag{4.3.3}$$

In this section, we will show how $\Delta(x)$ can be estimated with exponent pairs.

LEMMA 4.4. *Let γ be defined by (4.3.2). If $y \geq 1$ then*

$$\sum_{m\leq y} \frac{1}{m} = \log y + \gamma - \frac{\psi(y)}{y} + O\left(\frac{1}{y^2}\right).$$

Proof. Using Riemann-Stieltjes integration, we get

$$\sum_{m\leq y} \frac{1}{m} = \int_{1-}^{y} \frac{1}{u} d[u] = \int_{1-}^{y} \frac{1}{u} d(u - \psi(u))$$
$$= \log y + \frac{1}{2} - \int_1^\infty \frac{\psi(u)}{u^2} du - \frac{\psi(y)}{y} + \int_y^\infty \frac{\psi(u)}{u^2} du.$$

Comparing this with (4.3.2) shows that

$$\frac{1}{2} - \int_1^\infty \frac{\psi(u)}{u^2} du = \gamma.$$

To analyze the remaining integral, we let

$$\psi_1(u) = \int_0^u \psi(w) dw.$$

Since $\psi_1(1) = 0$, ψ_1 is periodic with period 1 and is therefore bounded. Integrating by parts, we get

$$\int_y^\infty \frac{\psi(u)}{u^2} du = \int_y^\infty \frac{d\psi_1(u)}{u^2} = -\frac{\psi_1(y)}{y^2} + 2\int_y^\infty \frac{\psi_1(u)}{u^3} du \ll \frac{1}{y^2}.$$

THEOREM 4.5. *Let $\Delta(x)$ be as defined in (4.3.3). Then*

$$\Delta(x) = -2 \sum_{n \le x^{1/2}} \psi\left(\frac{x}{n}\right) + O(1).$$

Proof. We begin by writing

$$\begin{aligned}\sum_{n\le x} d(n) &= \sum_{mr \le x} 1 \\ &= \sum_{m\le x^{1/2}} \sum_{r \le x/m} 1 + \sum_{r \le x^{1/2}} \sum_{m \le x/r} 1 - \sum_{m \le x^{1/2}} \sum_{r \le x^{1/2}} 1 \\ &= \sum\nolimits_1 + \sum\nolimits_2 - \sum\nolimits_3,\end{aligned}$$

say. By symmetry, $\sum_1 = \sum_2$. By Lemma 4.3,

$$\begin{aligned}\sum\nolimits_1 &= \sum_{m \le x^{1/2}} \left[\frac{x}{m}\right] = \sum_{m \le x^{1/2}} \left\{\frac{x}{m} - \psi\left(\frac{x}{m}\right) - \frac{1}{2}\right\} \\ &= x\left(\frac{1}{2}\log x + \gamma\right) - x^{1/2}\psi(x^{1/2}) - \frac{1}{2}x^{1/2} - \sum_{m\le x^{1/2}} \psi\left(\frac{x}{m}\right) + O(1).\end{aligned}$$

Furthermore,

$$\sum\nolimits_3 = [x^{1/2}]^2 = x - 2x^{1/2}\psi(x^{1/2}) - x^{1/2} + O(1),$$

and the desired result follows.

THEOREM 4.6. *If (k,l) is an exponent pair, then*

$$\Delta(x) \ll x^{(k+l)/(2k+2)} \log x.$$

If $(k,l) \ne (\frac{1}{2}, \frac{1}{2})$, then the $\log x$ factor in the above may be removed. In particular,

$$\Delta(x) \ll x^{27/82}.$$

Proof. We begin by writing

$$\sum_{m \le x^{1/2}} \psi\left(\frac{x}{n}\right) = \sum_{1 \le j \le J} \sum_{n \in I_j} \psi\left(\frac{x}{n}\right),$$

where

$$I_j = \{n : 2^{-j}x^{1/2} < n \le 2^{-j+1}x^{1/2}\}.$$

Employing Lemma 4.3, we get

$$\sum_{m\le x^{1/2}} \psi\left(\frac{x}{n}\right) \ll x^{(k+l)/(2k+2)} \sum_{1\le j\le J} 2^{-j(l-k)} + \sum_{1\le j\le J} 2^{-2j}. \tag{4.3.4}$$

The second sum on the right hand side is $\ll 1$. If $(k,l)=(\frac{1}{2},\frac{1}{2})$ then (4.3.4) is

$$\ll x^{1/3}\log x.$$

If $(k,l)\ne(\frac{1}{2},\frac{1}{2})$ then $l>k$ and

$$\sum_{1\le j\le J} 2^{-j(l-k)}$$

converges. In this case, we get the estimate

$$\sum_{m\le x^{1/2}} \psi\left(\frac{x}{n}\right) \ll x^{(k+l)/(2k+2)}.$$

The desired result now follows from Lemma 4.4.

The exponent 27/82 follows from taking $(k,l)=BA^3B(0,1)=(\frac{11}{30},\frac{26}{30})$.

4.4 THE CIRCLE PROBLEM

Let $r(n)$ denote the number of ways of writing n as a sum of two integral squares. It is well-known (see for expample, Hardy and Wright(1979), Chapter 16.9-10) that

$$r(n)=4\sum_{d|n}\chi(d), \tag{4.4.1}$$

where χ is the non-trivial Dirichlet character mod4; i.e.

$$\chi(d)=\begin{cases} 1 & \text{if } d\equiv 1 \bmod 4;\\ -1 & \text{if } d\equiv 3 \bmod 4;\\ 0 & \text{otherwise.}\end{cases}$$

Let $R(x)$ be defined by the relation

$$\sum_{n\le x} r(n)=\pi x+R(x). \tag{4.4.2}$$

From the point of view of exponential sums, $R(x)$ behaves very much like $\Delta(x)$. Indeed, in Theorem 4.8, we shall prove an estimate for $R(x)$ that is identical to the estimate for $\Delta(x)$ that was obtained in Theorem 4.6.

To begin our analysis of $R(x)$, we write

$$\begin{aligned}\sum_{n\le x}\sum_{d|n}\chi(d) &= \sum_{md\le x}\chi(d)\\ &= \sum_{d\le x^{1/2}}\chi(d)\sum_{m\le x/d}1+\sum_{m\le x^{1/2}}\sum_{d\le x/m}\chi(d)-\sum_{m\le x^{1/2}}\sum_{d\le x^{1/2}}\chi(d) \qquad (4.4.3)\\ &= \sum\nolimits_1+\sum\nolimits_2-\sum\nolimits_3,\end{aligned}$$

say. To analyze these sums further, we need more information about sums involving $\chi(m)$.

LEMMA 4.7. *Let $S(y)$ be defined by*

$$S(y) = \sum_{m \le y} \chi(m).$$

Then

$$S(y) = \frac{1}{2} - \psi\left(\frac{y-1}{4}\right) + \psi\left(\frac{y-3}{4}\right),$$

and

$$\sum_{d \le y} \frac{\chi(d)}{d} = \frac{\pi}{4} + \frac{S(y) - 1/2}{y} + O\left(\frac{1}{y}\right).$$

Proof. To prove the first statement, we note that

$$S(y) = \sum_{\substack{1 \le n \le y \\ n \equiv 1 \bmod 4}} 1 - \sum_{\substack{1 \le n \le y \\ n \equiv 3 \bmod 4}} 1.$$

Therefore

$$\sum_{d \le y} \frac{\chi(d)}{d} = \frac{\pi}{4} - \sum_{d > y} \frac{\chi(d)}{d} = \frac{\pi}{4} \int_y^\infty \frac{dS(u)}{u}.$$

Using integration by parts, we get

$$\begin{aligned}
\int_y^\infty \frac{dS(u)}{u} &= -\frac{S(y)}{y} + \int_y^\infty \frac{S(u)}{u^2} du \\
&= \frac{1}{2} - \frac{S(y)}{y} + \int_y^\infty \left\{\psi\left(\frac{y-3}{4}\right) - \psi\left(\frac{y-1}{4}\right)\right\} \frac{du}{u^2}.
\end{aligned}$$

The last two integrals can be analyzed using integration by parts and the function

$$\psi_1(u) = \int_0^u \psi(w) dw.$$

As we noted in the previous section, ψ_1 is bounded; therefore

$$\int_y^\infty \psi\left(\frac{u-a}{4}\right) \frac{du}{u^2} = 4 \int_y^\infty \frac{1}{u^2} d\psi_1\left(\frac{u-a}{4}\right) \ll \frac{1}{y^2}$$

when $a = 1$ or $a = 3$. The proof follows upon combining the last four displayed statements.

THEOREM 4.8. *Let $R(x)$ be defined by (4.4.2). Then*

$$R(x) = 4 \sum_{d \le x^{1/2}} \left\{\psi\left(\frac{x}{4d+1}\right) - \psi\left(\frac{x}{4d+3}\right) + \psi\left(\frac{x-3}{4d}\right) - \psi\left(\frac{x-1}{4d}\right)\right\} + O(1).$$

If (k,l) is an exponent pair, then

$$R(x) \ll x^{(k+l)/(2k+2)} \log x,$$

and the $\log x$ factor may be removed if $(k,l) \neq (\frac{1}{2}, \frac{1}{2})$. In particular, $R(x) \ll x^{27/82}$.

Proof. Let $\sum_1, \sum_2$, and $\sum_3$ be as defined in (4.4.2). By Lemma 4.7,

$$\begin{aligned}
\sum\nolimits_1 &= \sum_{d \leq x^{1/2}} \chi(d) \left[\frac{x}{d}\right] \\
&= x \sum_{d \leq x^{1/2}} \frac{\chi(d)}{d} - \frac{1}{2} S(x^{1/2}) - \sum_{d \leq x^{1/2}} \chi(d) \psi\left(\frac{x}{d}\right) \\
&= \frac{\pi}{4} x + (S(x^{1/2}) - \frac{1}{2}) x^{1/2} - \sum_{d \leq x^{1/2}} \chi(d) \psi\left(\frac{x}{d}\right) + O(1).
\end{aligned}$$

Similarly,

$$\begin{aligned}
\sum\nolimits_2 &= \sum_{m \leq x^{1/2}} S\left(\frac{x}{m}\right) \\
&= \frac{1}{2} x^{1/2} - \sum_{m \leq x^{1/2}} \{\psi(\frac{x^{1/2}-1}{4m}) + \psi(\frac{x^{1/2}-3}{4m})\} + O(1)
\end{aligned}$$

Finally,

$$\sum\nolimits_3 = x^{1/2} S(x^{1/2}) + O(1).$$

Combining the estimates for $\sum_1, \sum_2$, and $\sum_3$ gives the result.

The second statement follows from the first by applying Lemma 4.3 and a splitting-up argument; the details are essentially the same as those in the proof of Theorem 4.6. The third statement follows upon taking $(k,l) = BA^3B(0,1) = (\frac{11}{30}, \frac{26}{30})$.

4.5 GAPS BETWEEN SQUAREFREE NUMBERS

A number is said to be squarefree if it is divisible by no square other than 1. In this section, we will obtain an upper bound for the gaps between squarefree numbers.

If $\mu(n)$ is the Möbius function, then $\mu^2(n)$ is the characteristic function for the squarefree numbers. A simple argument suffices to show that

$$\sum_{n \leq x} \mu^2(n) = \frac{6}{\pi^2} x + O(x^{1/2}).$$

It follows that there is some C such that if $x \geq 1$ and $h \geq Cx^{1/2}$ then there is a squarefree number n in the range $x - h \leq n \leq x$. In this section, we will improve the exponent $1/2$ to $2/9$.

LEMMA 4.9. *Suppose $1 \le h \le x/2$ and that $h^{1/2} < M \le x^{1/2}$. Then*

$$\sum_{x-h<n\le x} \mu^2(n) = \frac{6h}{\pi^2} + O(R\log x + S\log x + h^{1/2}),$$

where

$$R = \max_{N\le M} \max_{x-h\le y\le x} \max_{N_1\le 2N} \Big| \sum_{N\le n\le N_1} \psi\left(\frac{y}{n^2}\right)\Big|$$

and

$$S = \max_{N\le xM^{-2}} \max_{x-h\le y\le x} \max_{N_1\le 2N} \Big| \sum_{N\le n\le N_1} \psi\left(\frac{y^{1/2}}{n^{1/2}}\right)\Big|.$$

Proof. We begin by writing

$$\sum_{x-h<n\le x} \mu^2(n) = \sum_{x-h<n\le x} \sum_{d^2|n} \mu(d) = \sum_{d\le x^{1/2}} \mu(d)\left\{\left[\frac{x}{d^2}\right] - \left[\frac{x-h}{d^2}\right]\right\}.$$

Let T be a parameter to be chosen later, and write the right-hand side of the above as

$$h\sum_{d\le T} \frac{\mu(d)}{d^2} + O(T) + O\Big(\sum_{T<d} \sum_{\substack{m \\ x-h<md^2\le x}} 1\Big).$$

The first term is

$$\frac{6}{\pi^2}h + O\left(\frac{h}{T}\right),$$

and the sum in the second error term is at most

$$\begin{aligned}
&\sum_{T<d\le M} \left\{\left[\frac{x}{d^2}\right] - \left[\frac{x-h}{d^2}\right]\right\} + \sum_{m\le xM^{-2}} \left\{\left[\frac{x^{1/2}}{m^{1/2}}\right] - \left[\frac{(x-h)^{1/2}}{m^{1/2}}\right]\right\} \\
&\ll \sum_{T<d} \frac{h}{d^2} + \sum_{m\le xM^{-2}} \frac{h}{x^{1/2}m^{1/2}} + \\
&\quad \Big|\sum_{d\le M} \psi\left(\frac{x}{d^2}\right) - \psi\left(\frac{x-h}{d^2}\right)\Big| + \Big|\sum_{m\le xM^{-2}} \psi\left(\frac{x^{1/2}}{m^{1/2}}\right) - \psi\left(\frac{(x-h)^{1/2}}{m^{1/2}}\right)\Big|.
\end{aligned}$$

After splitting up the third and fourth sums into $\ll \log x$ subsums, we find that the above is

$$\ll hT^{-1} + hM^{-1} + R\log x + S\log x.$$

Taking $T = h^{1/2}$ gives the result.

THEOREM 4.10. *There is some C such that if $x \geq 1$ and $h \geq Cx^{2/9} \log x$, then there is a squarefree number n such that $x - h < n \leq x$.*

Proof. From Lemma 4.3 with $(k, l) = (\frac{1}{2}, \frac{1}{2})$, we see that

$$\Big| \sum_{N \leq n \leq N_1} \psi(y^{1/2} n^{-1/2}) \Big| \ll x^{1/3} M^{-1/3} + x M^{-3}$$

whenever $N \leq xM^{-2}$, $N_1 \leq 2N$, and $x - h \leq y \leq x$. Similarly if (k, l) is an exponent pair, then

$$\Big| \sum_{N \leq n \leq N_1} \psi(yn^{-2}) \Big| \ll x^{k/(k+1)} N^{(l-2k)/(k+1)} + x^{-1} M^3$$

whenever $x - h \leq y \leq x$, $N_1 \leq 2N \leq 2x^{1/2}$, and $1 \leq h \leq M$. The theorem now follows by taking $(k, l) = (\frac{2}{7}, \frac{4}{7})$, $M = x^{2/5}$, and applying Lemma 4.9.

4.6 THE PIATETSKI-SHAPIRO PRIME NUMBER THEOREM

Let c be a positive real number, and let $\pi_c(x)$ denote the number of n not exceeding x for which $[n^c]$ is prime. When $c \leq 1$, the estimate

$$\pi_c(x) \sim \frac{x}{c \log x} \tag{4.6.1}$$

is an easy consequence of the Prime Number Theorem. On the other hand, (4.6.1) is not valid for $c = 2$ since n^2 is never prime. It is likely that (4.6.1) is true for $c < 2$. In this section we will prove Piatetski-Shapiro's (1953) original result on this problem; namely (4.6.1) is true for $c < 12/11$.

Let $\gamma = 1/c$. Then $[n^c]$ is prime if and only if there is a prime p such that

$$-(p+1)^\gamma \leq -n^\gamma < -p^\gamma.$$

Moreover, $n \leq x$ precisely when $p \leq x^c$, except possibly if $p \leq x^c \leq p + 1$. Therefore

$$\pi_c(x) = \sum_{p \leq x^c} \{[-p^\gamma] - [-(p+1)^\gamma]\} + O(1) = \sum\nolimits_1 + \sum\nolimits_2,$$

where

$$\sum\nolimits_1 = \sum_{p \leq x^c} ((p+1)^\gamma - p^\gamma),$$

and

$$\sum\nolimits_2 = \sum_{p \leq x^c} \{\psi(-(p+1)^\gamma) - \psi(-p^\gamma)\}.$$

The first sum can be estimated with Riemann-Stieltjes integration and the Prime Number Theorem. Let $\pi(x) = \pi_1(x)$. Then

$$\sum\nolimits_1 = \sum_{p \le x^c} \{\gamma p^{\gamma-1} + O(p^{\gamma-2})\} = \int_2^{x^c} \gamma u^{\gamma-1} d\pi(u) + O(x^{1-c} + 1).$$

The above integral is

$$\gamma x^{1-c}\pi(x^c) - \int_2^{x^c} \gamma(\gamma-1)u^{\gamma-2}\pi(u)\, du.$$

Using the Prime Number Theorem in the form

$$\pi(u) = \frac{u}{\log u} + O(\frac{u}{\log^2 u}),$$

we find that

$$\sigma_1 = \frac{x}{c \log x} + O(\frac{x}{\log^2 x}).$$

Note that from the trivial estimate $|\psi(w)| \le 1/2$ we get $\sum_2 \ll x^c$. This is sufficient to prove (4.6.1) when $c \le 1$. When $c = 1$, (4.6.1) is just the Prime Number Theorem. We shall henceforth assume that $c > 1$.

We begin our analysis of $\sum_3$ by invoking Vaaler's theorem (Theorem A.6) on approximating ψ. Here, it is convenient to restate this result as saying that for any $J > 0$, there exist functions ψ^* and δ such that

$$\psi(w) = \psi^*(w) + O(\delta(w))$$

where

$$\psi^*(w) = \sum_{1 \le |j| \le J} a(j) e(jw),$$

$$\delta(w) = \sum_{|j| \le J} b(j) e(jw),$$

$$a(j) \ll |j|^{-1},\ b(j) \ll J^{-1}.$$

Moreover, the function δ is non-negative. Consequently,

$$\begin{aligned}\sum\nolimits_2 &= \sum_{p \le x^c} \{\psi^*(-(p+1)^\gamma) - \psi^*(-p^\gamma)\} + O(\sum_{n \le x^c} \{\delta(-(n+1)^\gamma) + \delta(-n^\gamma)\}) \\ &= \sum\nolimits_3 + O\left(\sum\nolimits_4\right),\end{aligned}$$

say.

We can dispense with $\sum_4$ immediately. Suppose that $N \le x^c$ and $N_1 \le 2N$. Using the exponent pair $(\frac{1}{2}, \frac{1}{2})$, we get

$$\begin{aligned}\sum_{N<n\le N_1} \delta(-n^\gamma) &\ll \frac{1}{J} \sum_{|j|\le J} \Big| \sum_{N<n\le N_1} e(jn^\gamma)\Big| \\ &\ll J^{-1}N + J^{-1} \sum_{1\le j\le J} (j^{1/2}N^{\gamma/2} + j^{-1}N^{1-\gamma}) \\ &\ll J^{-1}N + J^{1/2}N^{\gamma/2} + J^{-1}N^{1-\gamma}\log N.\end{aligned}$$

Taking $J = N^{1-\gamma}\log^2 N$ gives

$$\sum_{N<n\le N_1} \delta(-n^\gamma) \ll N^\gamma \log^{-2} N + N^{1/2}\log N.$$

The same estimate holds if n is replaced by $n+1$. Combining this with a splitting-up argument yields

$$\sum\nolimits_4 \ll \frac{x}{\log^2 x} + x^{c/2}\log N \ll \frac{x}{\log^2 x}$$

provided $c < 2$.

To complete the proof of (4.6.1) it suffices to show that $\sum_3 \ll x/\log^2 x$ for the appropriate values of c. Let Λ denote von Mangoldt's lambda-function; i.e.

$$\Lambda(n) = \begin{cases} \log n & \text{if } n = p^a \text{ for some prime } p; \\ 0 & \text{otherwise.} \end{cases}$$

A simple argument using integration by parts shows that if $N' \le 2N$ and F is bounded, then

$$\sum_{N<p<N'} F(p) \ll \frac{1}{\log N} \max_{N_1\le 2N} \Big| \sum_{N<n\le N_1} \Lambda(n)F(n)\Big| + N^{1/2}.$$

Therefore our desired estimate for σ_3 will follow if we can show that

$$\sum_{N<n\le N_1} \Lambda(n)\{\psi^*(-(n+1)^\gamma) - \psi^*(-n^\gamma)\} \ll N^\gamma \log^{-2} N \tag{4.6.2}$$

whenever $N \le x^c$ and $N_1 \le 2N$.

On writing

$$\phi_j(x) = 1 - e(j(x^\gamma - (x+1)^\gamma))$$

we see that the right-hand side of (4.6.2) is

$$\begin{aligned} &= \sum_{1\le|j|\le J} a(j) \sum_{N\le n\le N_1} \Lambda(n)\{e(-jn^\gamma) - e(-j(n+1)^\gamma)\} \\ &= \sum_{1\le|j|\le J} a(j) \sum_{N\le n\le N_1} \Lambda(n)\phi_j(n)e(-jn^\gamma).\end{aligned}$$

By partial summation, the above is

$$\begin{aligned}
&\ll \sum_{1\le j\le J} j^{-1} \Big| \sum_{N<n\le N_1} \Lambda(n)\phi_j(n)e(-jn^\gamma)\Big| \\
&\ll \sum_{1\le j\le J} j^{-1} |\phi_j(N_1) \sum_{N<n\le N_1} \Lambda(n)e(jn^\gamma)| \\
&+ \int_N^{N_1} \sum_{1\le j\le J} j^{-1} \Big|\frac{\partial \phi_j(x)}{\partial x} \sum_{N\le n\le x} \Lambda(n)e(jn^\gamma)\Big| dx \\
&\ll N^{\gamma-1} \max_{N_2} \sum_{1\le j\le J} \Big| \sum_{N\le n\le N_2} \Lambda(n)e(jn^\gamma)\Big|,
\end{aligned}$$

where $N_2 \le 2N$ and $N \le x^c$. Here we use the bounds

$$\phi_j(x) \ll jN^{\gamma-1}, \ \frac{\partial \phi_j(x)}{\partial x} \ll jN^{\gamma-2}.$$

This completes the initial stage of our analysis. We state our results so far as

THEOREM 4.11. *If $c < 2$ then*

$$\pi_c(x) = \frac{x}{\log x} + O(\frac{x}{\log^2 x}) + O(S),$$

where

$$S = \max_{N\le x^c} \max_{N_2\le 2N} N^{\gamma-1} \sum_{1\le j\le J} \Big| \sum_{N\le n\le N_2} \Lambda(n)e(jn^\gamma)\Big|,$$

and $J = J(N) = N^{1-\gamma}\log^2 N$.

To treat the sum$\sum_3$ we need a way to convert sums involving $\Lambda(n)$ into more conventional sums. In the next lemma we present Vaughan's (1980) elegant method for making such conversions.

LEMMA 4.12. *(Vaughan) Let u and v be positive real numbers. If $n > v$ then*

$$\Lambda(n) = -\sum_{\substack{kl=n \\ k>v,l>u}} \Lambda(k) \sum_{\substack{d|l \\ d\le u}} \mu(d) + \sum_{\substack{kl=n \\ l\le u}} \mu(k)\log l - \sum_{\substack{klm=n \\ l\le u, m\le v}} \Lambda(k)\mu(m).$$

Proof. Let F and M denote the Dirichlet series

$$F(s) = \sum_{n\le v} \Lambda(n)n^{-s}, \ M(s) = \sum_{n\le u} \mu(n)n^{-s}.$$

The lemma follows from comparing coefficients of n^{-s} on both sides of the identity

$$\frac{\zeta'}{\zeta} + F = (\frac{\zeta'}{\zeta} + F)(1-\zeta M) - \zeta' M + \zeta F M.$$

Now assume that $N \geq v$. After multiplying both sides of the equation in Lemma 4.12 by $e(jn^\gamma)$ and summing over n, we get

$$\begin{aligned}
&\sum_{1\leq j\leq J}\sum_{N<n\leq N_1} \Lambda(n)e(jn^\gamma) \\
&= -\sum_{1\leq j\leq J}\sum_{u<m\leq N_1/v}\sum_{\substack{N/m<n\leq N_1/m \\ v<n}} a(m)\Lambda(n)e(jm^\gamma n^\gamma) \\
&\quad + \sum_{1\leq j\leq J}\sum_{m\leq u}\sum_{N/m<n\leq N_1/m} \mu(m)\log(n)e(jm^\gamma n^\gamma) \\
&\quad - \sum_{1\leq j\leq J}\sum_{m\leq uv}\sum_{N/m<n\leq N_1/m} b(m)e(jm^\gamma n^\gamma) \\
&= -S_1 + S_2 + S_3,
\end{aligned}$$

say. Here, the functions a and b are defined by

$$a(m) = \sum_{\substack{d|m \\ d\leq u}} \mu(d)\,, \; b(n) = \sum_{\substack{de=n \\ d\leq u, e\leq v}} \mu(d)\Lambda(e).$$

Note that $|a(m)| \leq d(m)$, and

$$|b(m)| \leq \sum_{d|m} \Lambda(d) = \log m.$$

Moreover,

$$\sum_{X<m\leq 2X} |a(m)|^2 \ll X\log^3 X\,, \quad \sum_{Y<n\leq 2Y} |b(m)|^2 \ll Y\log^2 Y.$$

In the nomenclature of Vaughan, sums of the form

$$\sum_{X<m\leq 2X}\sum_{N/m<n\leq N_1/m} \alpha(m)F(mn)$$

are called Type I sums, and sums of the form

$$\sum_{X<m\leq 2X}\sum_{Y<n\leq 2Y} \alpha(m)\beta(n)F(mn)$$

are called Type II sums. The sumS_1 can be split up into subsums of Type II, and S_3 can be split up into subsums of Type I. We will treat S_2 by splitting it into two different groups of subsums; the first group will be treated as Type I, and the second group will be treated as Type II. Our Type I estimates will be done very simply. The Type II estimates are more complicated; we present them in the following lemma.

LEMMA 4.13. *Suppose $\alpha(m)$ and $\beta(n)$ are sequences supported on the intervals $(X, 2X]$ and $(Y, 2Y]$ respectively. Suppose further that*

$$\sum_m |\alpha(m)|^2 \ll X \log^{2A} X, \sum_n |\beta(m)|^2 \ll Y \log^{2B} Y.$$

Let (k, l) be an exponent pair, and suppose that

$$(\kappa, \lambda) = A(k, l) = (k/(2k+2), (k+l+1)/(2k+2)).$$

Let j be a positive real number, and set $F = jX^\gamma Y^\gamma$. Finally, assume that $N_1 \leq 2N$. Then

$$\sum_m \sum_n \alpha(m)\beta(n)e(jm^\gamma n^\gamma) \ll (F^\kappa X^\lambda Y^{1-\kappa} + F^{-1/2}N + XY^{1/2}) \log^{A+B+1} N. \quad (4.6.3)$$

Proof. Let S denote the left-hand side of (4.6.3). We may normalize to the case $A = 0$ and $B = 0$. We may also assume that $XY/2 \leq N \leq 4XY$, for otherwise S is an empty sum. By Cauchy's inequality,

$$|S|^2 \ll X \sum_m \Big| \sum_{\substack{n \\ N < mn \leq N_1}} b(n)e(jm^\gamma n^\gamma)\Big|^2.$$

Now let Q be a parameter to be chosen later. We assume that $Q \leq Y$, and we apply the Weyl-van der Corput inequality (Lemma 2.5) to get

$$|S|^2 \ll \frac{X^2Y^2}{Q} + \frac{XY}{Q} \sum_{1 \leq |q| \leq Q} |\beta(n)\overline{\beta(n+q)}S_0(q;n)|$$

where

$$S_0(q;n) = \sum_{m \in I(q;n)} e(jm^\gamma(n^\gamma - (n+q)^\gamma)),$$

and $I(q;n)$ is a sub-interval of $(X, 2X]$. By Cauchy's inequality,

$$|b(n)\overline{b(n+q)}| \leq \frac{1}{2}(|b(n)|^2 + |b(n+q)|^2).$$

Moreover, $|S_0(q;n)| = |S_0(-q;n+q)|$, so we have

$$|S|^2 \ll \frac{X^2Y^2}{Q} + \frac{XY}{Q} \sum_n |b(n)|^2 \sum_{1 \leq q \leq Q} |S_0(q;n)|.$$

Upon using the exponent pair (k, l) , we get

$$\begin{aligned}\frac{1}{Q} \sum_{1 \leq q \leq Q} |S_0(q;n)| &\ll \frac{1}{Q} \sum_{1 \leq q \leq Q} \{(qFX^{-1}Y^{-1})^k X^l + (qF)^{-1}XY\} \\ &\ll Q^k F^k X^{l-k} Y^{-k} + F^{-1} N \log N.\end{aligned}$$

We therefore obtain

$$|S|^2 \ll (Q^{-1}X^2Y^2 + F^k Q^k X^{-k+l+1} Y^{2-k} + F^{-1}N) \log N.$$

The lemma now follows by appealing to Lemma 2.3.

We are now ready to prove Piatetski-Shapiro's theorem on π_c.

THEOREM 4.14. *Let (k, l) be an exponent pair. If*

$$c < \frac{4k+4}{4k+l+3}$$

then

$$\pi_c(x) \sim \frac{x}{c \log x}. \tag{4.6.4}$$

In particular, (4.6.4) is true if $c < 12/11$.

Proof. From Theorem 4.11 and the definitions of S_ν, we see that it suffices to show that $S_\nu \ll N \log^{-2} N$ for $\nu = 1, 2$, and 3.

We begin by considering S_2. Recall that

$$S_2 = \sum_{1 \le j \le J} \sum_{m \le u} \sum_{N/m < n \le N_1/m} \mu(m) \log(n) e(jm^\gamma n^\gamma).$$

Upon using the exponent pair $(\frac{1}{2}, \frac{1}{2})$ and partial summation, we find that

$$S_2 \ll \sum_{1 \le j \le J} \sum_{m \le u} \{(jN^\gamma)^{1/2} + (jm)^{-1} N^{1-\gamma}\} \log N \ll (uJ^{3/2} N^{\gamma/2} + N^{1-\gamma}) \log^3 N.$$

We now take $u = N^{\gamma - 1/2} \log^{-8} N$, and this gives the desired estimate for S_2.

The sumS_3 may be split into $\ll \log^2 N$ sums of the form

$$\sum_{1 \le j \le J} \sum_{X \le m \le 2X} \sum_{\substack{Y \le n \le 2Y \\ N_1 < mn \le N_1}} \alpha(m) \beta(m) e(jm^\gamma n^\gamma).$$

By exploiting the symmetry between m and n, we may arrange these sums so that $u < Y \le N^{1/2}$ and $N^{1/2} \le X \le \max(N/u, N/v)$. To simplify the latter condition, we assume that $v = u$. By Lemma 4.12, each subsum is

$$\ll \sum_{1 \le j \le J} \{(jN^\gamma)^\kappa X^\lambda Y^{1-\kappa} + j^{-1/2} N^{1-\gamma/2} + X^{1/2} Y\} \log^4 N$$
$$\ll \{N^{2-\gamma} X^{\lambda+\kappa-1} + N^{3/2-\gamma} + NX^{-1/2}\} \log^4 N,$$

where $(\kappa, \lambda) = A(k, l)$. After adding the subsums, we find that

$$S_3 \ll \{N^{3/2-\gamma+\lambda/2+\kappa/2} + N^{3/2-\gamma} + Nu^{-1/2}\} \log^{10} N.$$

To make the first term in the above expression sufficiently small, we need to assume that

$$\gamma < \frac{\lambda + \kappa + 1}{2} = \frac{4k+l+3}{4k+4}.$$

Note that since $\gamma = 1/c$, this is equivalent to our hypothesized condition on (k, l). For the other two terms, we require only that $\gamma > 1/2$, and this is a less restrictive condition.

We rewrite S_3 as

$$\sum_{m \le u} + \sum_{u < m \le uv} = S_4 + S_5.$$

By using essentially argument that was used for S_2, we find that

$$S_4 \ll (uJ^{3/2}N^{\gamma/2} + N^{1-\gamma}) \log^3 N.$$

For S_5, we repeat the argument used for S_1 to get

$$S_5 \ll \{N^{3/2-\gamma+\lambda/2+\kappa/2} + N^{3/2-\gamma} + Nu^{-1/2} + N^{1/2}u\} \log^{10} N.$$

The desired estimate for S_5 follows from the condition $\gamma > (\lambda + \kappa + 1)/2$.

The exponent 12/11 follows by taking the exponent pair $(\frac{1}{2}, \frac{1}{2})$.

4.7 NOTES

The most recent results on the zeta-function, divisor, and circle problems are reported in the notes to Chapter 7.

Theorem 4.2 is due to Phillips (1933).

The material in Section 3 is adapted from Richert (1953). The exponent for gaps between squarefree numbers was improved to $\theta = 0.221982\ldots$ by Rankin (1955); we will give this result in Chapter 5. The best known result at this time is $\theta = 3/14 = 0.2142857\ldots$, and is due to Filaseta and Trifonov (in press). Exponential sums play only a small role in their argument; only the exponent pair $(1/2, 1/2)$ is used. Most of the improvement comes from some clever combinatorial ideas.

Kolesnik (1969) improved the exponent in the Piatetski-Shapiro problem to 10/9. Heath-Brown (1983) sharpened the exponent to 755/662; his proof uses a more elaborate version of Vaughan's identity. Kolesnik (1985b) improved the exponent to 39/34.

5. COMPUTING OPTIMAL EXPONENT PAIRS

5.1 INTRODUCTION

In the last chapter, we saw several applications of exponent pairs where the ultimate result was expressed as a function of

$$\theta(k,l) = \frac{ak + bl + c}{dk + el + f} \tag{5.1.1}$$

where $a, b, \ldots, f$ are real numbers. In this chapter, we give an algorithm for computing $\inf \theta$, where the infimum is taken over all exponent pairs produced by the A- and B-processes.

Let $\mathbf{P}$ denote the set of all exponent pairs produced from $(0,1)$ by the A- and B-processes. Since B^2 is the identity, we may write $\mathbf{P} = A\mathbf{P} \cup BA\mathbf{P}$, where

$$A\mathbf{P} = \{A(k,l) : (k,l) \in \mathbf{P}\} \text{ and } BA\mathbf{P} = \{BA(k,l) : (k,l) \in \mathbf{P}\}.$$

Note that $(0,1) \in A\mathbf{P}$ since $A(0,1) = (0,1)$. The other exponent pairs in $A\mathbf{P}$ have the form

$$A^{q_1}BA^{q_2}B\ldots A^{q_k}B(0,1).$$

Similarly, the exponent pairs in $BA\mathbf{P}$ have the form

$$BA^{q_1}BA^{q_2}B\ldots A^{q_k}B(0,1).$$

The A and B processes may be regarded as linear transformations on projective space. Let

$$A = \begin{bmatrix} 1 & 0 & 0 \\ 1 & 1 & 1 \\ 2 & 0 & 2 \end{bmatrix}, B = \begin{bmatrix} 0 & 2 & -1 \\ 2 & 0 & 1 \\ 0 & 0 & 2 \end{bmatrix}.$$

Then

$$A\begin{bmatrix} k \\ l \\ 1 \end{bmatrix} = \begin{bmatrix} k \\ k+l+1 \\ 2k+2 \end{bmatrix}$$

and in projective space, this is equivalent to

$$\begin{bmatrix} k/(2k+2) \\ (k+l+1)/(2k+2) \\ 1 \end{bmatrix} = \begin{bmatrix} \kappa \\ \lambda \\ 1 \end{bmatrix},$$

where $(\kappa, \lambda) = A(k, l)$. A similar relation holds for the B matrix. We are of course, abusing notation by using the letters A and B in two different senses, but the precise meaning should be clear from context. The matrix representations are particularly useful for computing compositions. For example, the matrix representing the composition $A(B(k, l))$ is simply AB.

Let θ be as defined in (5.1.1). The notation $\inf \theta$ will be understood to mean

$$\inf\{\theta(k, l) : (k, l) \in \mathbf{P}\}.$$

Similarly, if $\mathbf{Q} \subseteq \mathbf{P}$, then

$$\inf_{\mathbf{Q}} \theta = \inf\{\theta(k, l) : (k, l) \in \mathbf{Q}\}.$$

By Lemma 2.3, exponent pairs have a convexity property; that is, if (k_1, l_1) and (k_2, l_2) are exponent pairs then so is $(tk_1 + (1-t)k_2, tl_1 + (1-t)l_2)$ for any t with $0 \le t \le 1$. Thus the convex hull of $\mathbf{P}$, which we will denote by conv $\mathbf{P}$, is a set of exponent pairs.

Throughout most of this chapter, we will be working with $\mathbf{P}$ instead of conv $\mathbf{P}$. One reason for this is that for the functions θ that are considered here,

$$\inf_{\mathbf{P}} \theta = \inf_{\text{conv } \mathbf{P}} \theta$$

(see Lemma 5.4). Moreover, we could state an analog of Lemma 5.1 for conv $\mathbf{P}$, but it would be more complicated. On the other hand, certain arguments can be made simpler by appropriate use of convexity; two such arguments are Applications 3 and 4 of Section 5.4.

5.2 PRELIMINARY LEMMAS

Our first lemma collects together several simple facts about the geometry of $\mathbf{P}$.

LEMMA 5.1.

(a) If $(k, l) \in \mathbf{P}$ then $k + l \le 1$.

(b) If $(k, l) \in A\mathbf{P}$ then $l - k - 1/2 \ge 0$.

(c) If $(k, l) \in BA\mathbf{P}$ then $l - k - 1/2 \le 0$.

(d) $A\mathbf{P} \cap BA\mathbf{P} = \{(1/6, 2/3)\}$.

Proof. If $(k, l) \in \mathbf{P}$ then either $(k, l) = A(\kappa, \lambda)$ or $(k, l) = BA(\kappa, \lambda)$ for some $(\kappa, \lambda) \in \mathbf{P}$. In either case,

$$k + l = \frac{2\kappa + \lambda + 1}{2\kappa + 2} \le 1.$$

This proves (a). To prove (b), suppose $(k,l) = A(\kappa,\lambda)$. Then

$$l - k - 1/2 = \frac{\lambda - \kappa}{2\kappa + 2} \geq 0 \tag{5.2.1}$$

since $\lambda \geq 1/2 \geq \kappa$. Part (c) follows from (a) and (b) and the observation that B is a reflection through the line $l - k - 1/2$. Part (d) follows from (b),(c), and the observation that equality holds in (5.2.1) if and only if $(\kappa,\lambda) = (1/2,1/2)$. Note that $(1/6,2/3) = AB(0,1) = BAB(0,1)$.

In geometric terms, (a) states that $\mathbf{P}$ lies in the triangle with vertices $(0,1)$, $(0,1/2)$, and $(1/2,1/2)$. In fact, it is easy to show that for any $q \geq 0$, $A^q\mathbf{P}$ lies in the triangle with vertices $(0,1)$, $(0,1/2)$, and $A^q(1/2,1/2)$.

We now give three lemmas on θ. We shall need to assume that

$$dk + el + f > 0 \tag{5.2.2}$$

for $(k,l) \in \mathbf{P}$. In practice, it is usually easier to establish that (5.2.2) holds for $(k,l) = (0,1/2),(0,1)$, and $(1/2,1/2)$. This is sufficient, for if (5.2.2) holds for these three points, it must also hold in the convex hull of these points.

We may regard θ as a 2×3 matrix, that is,

$$\theta = \begin{bmatrix} a & b & c \\ d & e & f \end{bmatrix}.$$

We denote the three subdeterminants of θ by

$$u = bf - ce, v = af - cd, \text{ and } w = ae - bd. \tag{5.2.3}$$

Furthermore, let

$$\xi(\theta) = \begin{bmatrix} u \\ v \\ w \end{bmatrix}.$$

We shall see later that computing $\inf\theta$ depends only on the projective equivalence class of $\xi(\theta)$.

LEMMA 5.2. *Suppose that θ is defined by (5.1.1) and let u, v, and w be defined by (5.2.3). Then for any ordered pairs (k,l) and (κ,λ),*

$$(\theta(k,l) - \theta(\kappa,\lambda))(dk + el + f)(d\kappa + e\lambda + f) = \det\begin{bmatrix} k & l & 1 \\ \kappa & \lambda & 1 \\ u & -v & w \end{bmatrix}.$$

The proof is a straightforward computation, and it is left to the reader. This lemma is actually a special case of the Cauchy-Binet formula.

LEMMA 5.3. *Let θ, u, v, and w be as in the previous lemma. Suppose also that θ satisfies (5.2.2). If $(k,l) \in \mathbf{AP}$ and $(k,l) \neq (1/6, 2/3)$ then $\theta B(k,l) - \theta(k,l)$ has the same sign as $w(k+l)+v-u$.*

Proof. By Lemma 5.2 and (5.2.2), $\theta B(k,l) - \theta(k,l)$ has the same sign as

$$\det \begin{bmatrix} l-1/2 & k+1/2 & 1 \\ k & l & 1 \\ u & -v & w \end{bmatrix}.$$

A few calculations with row operations show that this is equal to $(l-k-1/2) \times (w(k+l)+v-u)$. The first factor does not affect the sign since it is positive by parts (b), (c), and (d) of Lemma 5.1.

LEMMA 5.4. *Let θ be as in Lemma 5.2. Assume that (k_0, l_0) and (k_1, l_1) are exponent pairs and that*

$$\theta(k_0, l_0) \leq \theta(k_1, l_1).$$

For t with $0 \leq t \leq 1$, define

$$(k_t, l_t) = (1-t)(k_0, l_0) + t(k_1, l_1).$$

Then

$$\theta(k_0, l_0) \leq \theta(k_t, l_t) \leq \theta(k_1, l_1).$$

Proof. From Lemma 5.2, we see that $\theta(k_t, l_t) - \theta(k_0, l_0)$ has the same sign as

$$\det \begin{bmatrix} k_t & l_t & 1 \\ k_0 & l_0 & 1 \\ u & -v & w \end{bmatrix} = t \det \begin{bmatrix} k_1 & l_1 & 1 \\ k_0 & l_0 & 1 \\ u & -v & w \end{bmatrix}.$$

The right-hand side is non-negative by hypothesis, so $\theta(k_t, l_t) \geq \theta(k_0, l_0)$. A similar argument shows that $\theta(k_1, l_1) \geq \theta(k_t, l_t)$.

5.3 THE ALGORITHM

Since $\mathbf{P} = A\mathbf{P} \cup BA\mathbf{P}$, it follows that either $\inf \theta = \inf \theta A$ or $\inf \theta = \inf \theta BA$. Our next theorem decides which option is true in most circumstances.

THEOREM 5.5. *Let θ, defined by (5.1.1), be a function satisfying (5.2.2). Let u, v, and w be defined by (5.2.3). Let r be any real number such that*

$$r \leq \inf (k+l).$$

Define

$$Y = \max(wr + v - u, w + v - u)$$

and

$$Z = \min(wr + v - u, w + v - u).$$

If $Z \geq 0$, *then* $\inf \theta = \inf \theta A$. *If* $Y \leq 0$, *then* $\inf \theta = \inf \theta BA$.

Note that Y and Z depend only on $\xi(\theta)$. Moreover, if $\xi(\theta)$ is multiplied by a positive constant, the signs of Y and Z do not change. This justifies our earlier remark that $\inf \theta$ depends only on the projective equivalence class of θ.

Theorem 5.5 gives no information when $Z < 0 < Y$. We shall refer to this as the *indeterminate case.* In this instance, we resort to Theorem 5.6.

Proof. If $(k,l) \in \mathbf{AP}$, then $Z \leq w(k+l) + v - u \leq Y$. Assume that $Z \geq 0$. By Lemma 5.3, we see that $\theta B(k,l) \geq \theta(k,l)$ for $(k,l) \in \mathbf{AP}$. Consequently,

$$\inf_{B\mathbf{AP}} \theta \geq \inf_{\mathbf{AP}} \theta.$$

Since $\mathbf{P} = \mathbf{AP} \cup B\mathbf{AP}$, we see that $\inf \theta = \inf \theta A$.

On the other hand, if $Y \leq 0$ then $\theta B(k,l) \leq \theta(k,l)$ for all $(k,l) \in \mathbf{AP}$, and so $\inf \theta = \inf \theta BA$.

To apply Theorem 5.5, it is necessary to have a good lower bound for $\inf k + l$. The lower bound $r = 1/2$ follows from the definition of exponent pairs. We can use Theorem 5.5 with this value of r and

$$\theta = \begin{bmatrix} 1 & 1 & 0 \\ 0 & 0 & 1 \end{bmatrix}.$$

We find that $Y = Z = 0$, so $\inf \theta = \inf \theta A = \inf \theta BA$. We arbitrarily choose θA, and we set

$$\theta_1 = \theta A = \begin{bmatrix} 2 & 1 & 1 \\ 2 & 0 & 2 \end{bmatrix}.$$

Since $\xi(\theta_1) = [1 \quad 1 \quad -1]^T$, we get $Y = -1/2$ and $\inf \theta = \inf \theta ABA$. Now

$$\theta ABA = \begin{bmatrix} 4 & 2 & 3 \\ 4 & 2 & 4 \end{bmatrix}.$$

On the triangle with vertices $(0,1), (0,1/2)$, and $(1/2,1/2)$, the function $(4k + 2l + 3)/(4k + 2l + 4)$ assumes its minimum at $(0,1/2)$. Therefore $\inf k + l \geq 4/5$.

Continuing in this fashion, we find that

$$\begin{aligned}\inf k + l \geq & \theta ABA^3BA^2BABABA^2BABABA^2BA^2BABA^2BA^2BA^2BABA(0,1/2) \\ = & \frac{89037450}{107400671} > 0.8290213568591.\end{aligned}$$

The next result provides a means of analyzing the indeterminate case. It does not provide an answer for every possible θ in the indeterminate case, but we have found it to be sufficient for our applications.

THEOREM 5.6. *Let θ and r be as in Theorem 5.5. Let C be some finite product of A's and B's such that*

$$\inf \theta BA = \inf \theta BAC.$$

Suppose further that

$$\sup\{k+l : (k,l) \in C A \mathbf{P}\} = r_1.$$

If $\min(rw+v-u, r_1 w+v-u) \geq 0$, then $\inf \theta = \inf \theta A$.

Proof. Assume $(k,l) \in BAC\mathbf{P}$. By hypothesis, $k+l \leq r_1$. This together with our other hypotheses and Lemma 5.3 yields $\theta B(k,l) \geq \theta(k,l)$. Consequently,

$$\inf_{\mathbf{P}} \theta \leq \inf_{AC\mathbf{P}} \theta \leq \inf_{AC\mathbf{P}} \theta B = \inf_{\mathbf{P}} \theta BAC = \inf_{\mathbf{P}} \theta BA.$$

To illustrate how Theorem 5.6 can be used, consider the following example, which arises in the computation of $\inf k+l$. Suppose θ is such that

$$\xi(\theta) = [u \quad v \quad w]^T = [386 \quad 801 \quad -450]^T.$$

Now compute Z and Y from the formulas given Theorem 5.5. We find that $Z < 0 < Y$, so θ is in the indeterminate case. Now consider θBA. Note that

$$\xi(\theta BA) = [576 \quad 386 \quad -351]^T;$$

consequently, $\inf \theta BA = \inf \theta BABA$. Now we claim that if $(k,l) \in ABA\mathbf{P}$ then $k+l \leq 6/7$. For if $(k,l) = ABA(\kappa,\lambda)$, then $k+l = 1 - 1/(4\kappa+2\lambda+4)$, and $4\kappa+2\lambda+4 \leq 7$ by Lemma 5.1(a). Going back to $\xi(\theta)$, we find that

$$\min(rw+v-u, 6w/7+v-u) \geq 0,$$

so $\inf \theta = \inf \theta A$.

In the next corollary, we collect three special cases of Theorem 5.6. We omit the proofs since they are very similar to the discussion in the previous paragraph.

COROLLARY 5.7. *Let θ and r be as in Theorem 5.5. If any of the following hypotheses hold, then $\inf \theta = \inf \theta A$.*

(a) $\inf \theta BA = \inf \theta BABA$ and $\min(rw+v-u, 6w/7+v-u) \geq 0$.

(b) $\inf \theta BA = \inf \theta BABA^2$ and $\min(rw+v-u, 5w/6+v-u) \geq 0$.

(c) $\inf \theta BA = \inf \theta BABA^3BA$ and $\min(rw+v-u, 73w/88+v-u) \geq 0$.

The main part of our algorithm consists of examining $\xi(\theta)$ and using Theorem 5.6 or Corollary 5.7 to determine whether $\inf \theta = \inf \theta A$ or $\inf \theta = \inf \theta BA$. We then replace $\xi(\theta)$ by the appropriate choice of $\xi(\theta A)$ or $\xi(\theta BA)$, and we repeat the process. After an appropriate number of iterations, we expect to have

$$\inf \theta = \inf \theta B^j A^{q_1} BA^{q_2} B \ldots A^{q_r} BA,$$

where $j = 0$ or 1 and $q_1, q_2, \ldots, q_r$ are positive integers. However, it is possible to have

$$\inf \theta = \inf \theta A^q$$

for every $q \geq 0$. The next theorem gives us an easy way of recognizing when this happens.

THEOREM 5.8. *Let θ, u, v, and w be as in Theorem 5.5. The following conditions are equivalent.*

(a) $\inf \theta = \inf \theta A^q B$ for every $q \geq 0$.

(b) $\inf \theta = \theta(0,1)$.

(c) $w + v \geq u$ and $u \leq 0$.

Proof. We shall first prove that (a) and (b) are equivalent, and then we shall complete the proof by showing that (b) and (c) are equivalent.

Assume first that $\inf \theta = \inf \theta A^q$ for every $q \geq 0$. Then we can find a sequence $\{(k_q, l_q)\}$ of exponent pairs such that $(k_q, l_q) \in A^q \mathbf{P}$ and

$$\inf \theta = \lim_{q \to \infty} \theta(k_q, l_q).$$

Now we can show by induction that

$$A^q = \begin{bmatrix} 1 & 0 & 0 \\ Q - q - 1 & 1 & Q - 1 \\ 2Q - 2 & 0 & Q \end{bmatrix}, \tag{5.3.1}$$

where $Q = 2^q$. Therefore, if $(k_q, l_q) = A(k, l)$ then

$$(k_q, l_q) = (\frac{k}{(2Q-2)k + Q}, 1 - \frac{k + l - 1}{(2Q-2)k + Q}).$$

It follows that $k_q \leq 1/Q$ and $1 - l_q \leq (q+1)/2Q$; therefore

$$\lim_{q \to \infty} (k_q, l_q) = (0, 1).$$

Now θ is uniformly continuous on the convex hull of the triangle with vertices $(0, 1/2)$, $(0, 1)$, and $(1/2, 1/2)$, so

$$\lim_{q \to \infty} \theta(k_q, l_q) = \theta(0, 1).$$

Now assume that $\inf \theta = \theta(0,1)$. For any $q \geq 0$, $A^q \mathbf{P} \subseteq \mathbf{P}$, so $\inf \theta \leq \inf \theta A^q$. On the other hand, $(0,1) = A^q(0,1)$, so

$$\inf \theta A^q \leq \theta(0,1) = \inf \theta.$$

This completes the proof that (a) and (b) are equivalent.

Next, we prove that (b) and (c) are equivalent. By Lemma 5.2, $\theta(0,1) - \theta(k,l)$ has the same sign as

$$T = u(1 - l - k) - k(w + v - u).$$

If $w + v - u \geq 0$ and $u \leq 0$, then $T \leq 0$ for all $(k,l) \in \mathbf{P}$, so $\inf \theta = \theta(0,1)$. Conversely, assume that $\inf \theta = \theta(0,1)$. Then $T \leq 0$ for all $(k,l) \in \mathbf{P}$. In particular, $T \leq 0$ for $(k,l) = A^q B(0,1)$. From (5.3.1) we see that

$$A^q B(0,1) = (\frac{1}{4Q-2}, 1 - \frac{q+1}{4Q-2}),$$

where $Q = 2^q$. Therefore

$$qu \leq w + v - u \tag{5.3.2}$$

for all non-negative integers q. From this with $q = 0$, we see that $w + v - u \geq 0$. We also see that $u \leq 0$, for otherwise (5.3.2) would be false for q sufficiently large. This completes the proof.

The example $\theta(k, l) = k$ shows that it is possible to have $\inf \theta = \theta(0, 1)$. The example $\theta(k, l) = l$ shows that it is possible to have $\inf \theta = \theta(\frac{1}{2}, \frac{1}{2})$. The following theorem shows that this property holds for no other element of $\mathbf{P}$.

THEOREM 5.9. *Suppose θ satisfies the conditions of Theorem 5.5, and suppose there is some $(k, l) \in \mathbf{P}$ such that $\inf \theta = \theta(k, l)$. Then either $\inf \theta = \theta(0, 1)$ or $\inf \theta = \theta(1/2, 1/2)$.*

Proof. Suppose there is some $(k, l) \in \mathbf{P}$ such that $\inf \theta = \theta(k, l)$. Now (k, l) must have the form

$$B^j A^{q_1} BA \ldots A^{q_r} B(0, 1),$$

where $j = 0$ or 1 and the q_i's are positive integers. If $r = 0$ then $(k, l) = (1/2, 1/2)$ or $(0, 1)$, and we are done. Assume that $r \geq 1$, and let

$$\theta_1 = \theta B^j A^{q_1} BA \ldots A^{q_{r-1}} BA^{q_r - 1}.$$

Now $\inf \theta \leq \inf \theta_1$ since

$$B^j A^{q_1} BA \ldots A^{q_{r-1}} BA^{q_r - 1} \mathbf{P} \subseteq \mathbf{P}.$$

Furthermore, $\inf \theta = \theta_1 AB(0, 1) \geq \inf \theta_1$. Therefore, $\inf \theta_1 = \theta_1(\frac{1}{6}, \frac{2}{3})$.

Now $ABA^3B(0, 1) = (\frac{11}{82}, \frac{57}{82})$ and $BABA^3B(0, 1) = (\frac{8}{41}, \frac{26}{41})$. Consequently,

$$\theta_1(\frac{1}{6}, \frac{2}{3}) \leq \min\{\theta_1(\frac{11}{82}, \frac{57}{82}), \theta_1(\frac{8}{41}, \frac{26}{41})\} \leq \theta_1(\frac{27}{164}, \frac{109}{164}); \tag{5.3.3}$$

the last inequality following by an application of Lemma 5.4. In a similar fashion, we also obtain

$$\theta_1(\frac{1}{6}, \frac{2}{3}) \leq \min\{\theta_1(0, 1), \theta_1(\frac{1}{2}, \frac{1}{2})\} \leq \theta_1(\frac{1}{4}, \frac{3}{4}). \tag{5.3.4}$$

Now $(\frac{1}{6}, \frac{2}{3})$ lies on the line segment joining $(\frac{27}{164}, \frac{109}{164})$ and $(\frac{1}{4}, \frac{3}{4})$. Thus we may apply Lemma 5.4 again to get

$$\theta_1(\frac{1}{6}, \frac{2}{3}) \geq \min\{\theta_1(\frac{27}{164}, \frac{109}{164}), \theta_1(\frac{1}{4}, \frac{3}{4})\}. \tag{5.3.5}$$

From (5.3.3) through (5.3.5), we see that

$$\theta_1(\frac{1}{6}, \frac{2}{3}) = \min\{\theta_1(\frac{27}{164}, \frac{109}{164}), \theta_1(\frac{1}{4}, \frac{3}{4})\}.$$

From the proof of Lemma 5.4, we see that this can happen only if

$$\theta_1(\frac{27}{164}, \frac{109}{164}) = \theta_1(\frac{1}{6}, \frac{2}{3}) = \theta_1(\frac{1}{4}, \frac{3}{4}).$$

This together with (5.3.4) shows that

$$\theta_1(\frac{1}{4}, \frac{3}{4}) = \min\{\theta_1(0,1), \theta_1(\frac{1}{2}, \frac{1}{2})\}.$$

Using Lemma 5.4 again, we get

$$\theta_1(\frac{1}{6}, \frac{2}{3}) = \theta_1(\frac{1}{4}, \frac{3}{4}) = \theta(0,1) = \theta(\frac{1}{2}, \frac{1}{2});$$

therefore $\inf \theta_1 = \theta_1(0,1)$. Now

$$\theta_1(0,1) = \theta B^j A^{q_1} B \ldots A^{q_{r-1}} B A^{q_{r-1}}(0,1) = \theta B^j A^{q_1} B \ldots A^{q_{r-1}} B(0,1).$$

In other words, $\inf \theta = \theta(k_1, l_1)$, where

$$(k_1, l_1) = B^j A^{q_1} B \ldots A^{q_{r-1}} B(0,1).$$

Repeating this argument, we get $\inf \theta = \inf \theta(k_r, l_r)$, where $(k_r, l_r) = B^j(0,1)$ is either $(0,1)$ or $(1/2, 1/2)$. This completes the proof.

We can now give a step-by-step description of an algorithm for determining optimal exponent pairs.

1. Check θ for condition (5.2.2); that is, $dk + el + f \geq 0$.
2. Compute $\xi(\theta)$.
3. Use Theorem 5.8 to see if $\inf \theta = \theta(0,1)$. If so, then stop.
4. Use Theorem 5.8 on θB to see if $\inf \theta = \theta(\frac{1}{2}, \frac{1}{2})$. If so, then stop.
5. Use Theorem 5.5 to see if $\inf \theta = \theta A$ or $\inf \theta = \theta BA$. If this fails, try Theorem 5.6. If this fails, give up.
6. If $\inf \theta = \inf \theta A$, replace $\xi(\theta)$ by $\xi(\theta A)$. If $\inf \theta = \inf \theta BA$, replace $\xi(\theta)$ by $\xi(\theta BA)$. In either case, return to step 5.

Theorem 5.9 shows that is unnecessary to repeat steps 2 and 3 after the first loop. It is also unnecessary to repeat step 1 after the first loop. To see why, suppose that θ satisfies (5.2.2) and that $\inf \theta = \inf \theta A$, for example. Let

$$\theta A = \begin{bmatrix} a_1 & b_1 & c_1 \\ d_1 & e_1 & f_1 \end{bmatrix}.$$

Then

$$d_1 k + e_1 l + f = d\kappa + e\lambda + f,$$

where $(k, l) = A(\kappa, \lambda)$. The right hand side is positive since (5.2.2) is satisfied for θ.

5.4 APPLICATIONS

The results of the algorithm of the preceding section may be regarded as a sequence of exponent pairs that yield approximations to the desired infimum. The terms in this sequence are

$$A^{q_1}B(0,1)\,,\ A^{q_1}BA^{q_2}B(0,1)\,,$$

et cetera. We shall call the sequence of exponents $(q_1, q_2, q_3, \ldots)$ the q-sequence. Furthermore, we shall use score-by-innings notation and write the q-sequence as

$$q_1 q_2 q_3 \cdots$$

In this section, we will give the results that algorithm yields when applied to the problems discussed in Chapter 4. The computations reported here were done with MACSYMA.

Rankin's Constant. The quantity

$$\inf k + l$$

plays an important role in our algorithm. We call this quantity Rankin's Constant, and we denote it by the letter R. The first 50 terms in the q-sequence are

13211 21122 12221 21122 11213 21112 11132 11132 11221 11122.

From these terms, we find that

$$R = 0.829021356859133592409239777283112050988343270 3 \pm 8 \times 10^{-43}.$$

In some of the other applications discussed in Chapter 4, the optimal results can be expressed in terms of R. For example, let

$$\theta(k,l) = \frac{k+l}{2k+2}.$$

and let $\alpha = \inf \theta$. We see from Theorems 4.6 and 4.8 that

$$\Delta(x) \ll x^{\alpha+\epsilon} \text{ and } R(x) \ll x^{\alpha+\epsilon}$$

for every $\epsilon > 0$. Let $\theta_1(k,l) = k + l$. As we mentioned in the previous section, $\inf \theta_1 = \inf \theta_1 A$. Moreover,

$$\theta_1 A(k,l) = \frac{2k+l+1}{2k+2} = \theta(k,l) + \frac{1}{2}.$$

Therefore $\alpha = R - 1/2 = 0.32902135685913359240923977 72\ldots$.

In Theorem 4.13, we showed that if

$$\theta(k,l) = \frac{4k+l+3}{4k+4}$$

and $c < (\theta(k,l))^{-1}$, then

$$\pi_c \sim \frac{x}{c\log x}. \tag{5.4.1}$$

Again, let $\theta_1(k,l) = k + l$. Since

$$\theta(k,l) = \frac{1}{2}(\theta_1 A + 1)$$

we see that $\inf\theta = (R+1)/2$. Therefore, (5.4.1) is true for

$$c < (\inf\theta)^{-1} = 1.0934809440576886234567678933665528741229168\ldots.$$

Heath-Brown's improvement of Piatetski-Shapiro. Lest the reader obtain the impression that all applications of the algorithm reduce to computing R, we mention the following result due to Heath-Brown (Heath-Brown (1983)). If (k,l) is an exponent pair and

$$\theta(k,l) = \frac{12k+10}{12k-2l+13},$$

then $\pi_c \sim x/(c\log x)$ for $c \leq (\theta(k,l))^{-1}$. The first 50 terms of the q-sequence are

01222 11111 11121 11211 11121 12121 12123 21211 12211 11111.

It follows that (5.4.1) is true for

$$c < 1.1404913648148855613299352226496085 54\ldots.$$

Gaps between square-free numbers. Let

$$S(N) = \sum_{N<n\leq 2N} \psi(xn^{-2}).$$

From Lemma 4.3, we see that if (k,l) is an exponent pair and $N = x^\alpha$ then

$$S(N) \ll x^{\theta_\alpha(k,l)},$$

where

$$\theta_\alpha(k,l) = \frac{k+(l-2k)\alpha}{k+1}$$

For fixed α, let $g(\alpha) = \inf\theta_\alpha(k,l)$. From Lemma 4.9, we see that if

$$\delta > \sup\{g(\alpha) : 0 \leq \alpha \leq 1/2\},$$

then there is a square-free number n in the interval

$$x - x^{\delta} < n \le x$$

when x is sufficiently large.

We can use the algorithm of Section 5.3 to compute $g(\alpha)$ for any fixed α, but that algorithm gives us no direct method of computing $\sup g(\alpha)$. Instead, we use the following argument to show that

$$\sup g(\alpha) = \frac{2R+1}{6R+7}.$$

We begin by treating the range $\alpha < 3/10$. Let

$$(k,l) = BABABA^2BAB(0,1) = (\frac{13}{49}, \frac{57}{98}).$$

Since $l \ge 2k$, we see that

$$\theta\alpha(k,l) \le \theta_{\frac{3}{10}}(k,l) = \frac{55}{248} < \frac{(2R+1)}{(6R+7)}$$

when $\alpha \le 3/10$.

Now suppose $\alpha \ge 4/13$. If

$$(k,l) = BA^2BA^2BAB(0,1) = (\frac{19}{62}, \frac{52}{93})$$

then $l > 2k$, and so

$$\theta_\alpha(k,l) \le \theta_{\frac{4}{13}}(k,l) = \frac{701}{3159} < \frac{(2R+1)}{(6R+7)}.$$

We may therefore assume henceforth that $3/10 < \alpha < 4/13$.

We can use Theorem 5.5 to show that $\inf \theta_\alpha = \inf \theta_\alpha BA$ when $3/10 < \alpha < 4/13$. We therefore set

$$\psi_\alpha(k,l) = \theta_\alpha BA(k,l) = \frac{l + \alpha(2k - 2l + 1)}{2k + l + 2},$$

so that $g(\alpha) = \inf \psi_\alpha$.

We next show that $g(\alpha) \le (2R+1)/(6R+7)$ for all α under consideration. For (k,l) in $\mathbf{P}$, use convexity to construct the exponent pair

$$(k',l') = \frac{1}{2}((k,l) + B(k,l)) = \frac{1}{4}(2k + 2l - 1, 2k + 2l + 1).$$

By Lemma 5.4,

$$g(\alpha) \le \inf \psi_\alpha(k',l') = \inf \frac{2(k+l)+1}{6(k+l)+7} = \frac{2R+1}{6R+7}.$$

We now show that there is some β such that $g(\beta) \ge (2R+1)/(6R+7)$. Define $g_A = \inf \psi_\alpha A$ and $g_B = \inf \psi_\alpha BA$, so that $g(\alpha) = \min(g_A(\alpha), g_B(\alpha))$. From Lemma 5.1(b) we see that for fixed (k,l), $\psi_\alpha A(k,l)$ is an decreasing function of α. Therefore $g_A(\alpha)$ is decreasing. Similarly, $g_B(\alpha)$ is an increasing function. From Theorem 5.5, we see that $g_B(3/10) < g_A(3/10)$ and $g_B(4/13) > g_A(4/13)$. Now g_A and g_B are continuous, so there is some β for which $g_A(\beta) = g_B(\beta)$.

Now let $\{(k'_n, l'_n)\}$ and $\{(k''_n, l''_n)\}$ be sequences of exponent pairs (in $A\mathbf{P}$ and $BA\mathbf{P}$, respectively) such that

$$g(\beta) = \lim \psi_\beta(k'_n, l'_n) = \lim \psi_\beta(k''_n, l''_n).$$

For each n, we can find a convex combination (k_n, l_n) of (k'_n, l'_n) and (k''_n, l''_n) with the property that

$$k_n = l_n - \frac{1}{2}.$$

By Lemma 5.4, $\psi_\beta(k_n, l_n)$ lies in an interval whose endpoints are $\psi_\beta(k'_n, l'_n)$ and $\psi_\beta(k''_n, l''_n)$, whence

$$g(\beta) = \lim \psi_\beta(k_n, l_n).$$

If we let $r_n = k_n + l_n$, we see that

$$\psi_\beta(k_n, l_n) = (2r_n + 1)/(6r_n + 7),$$

so that

$$g(\beta) = lim(2r_n + 1)/(6r_n + 7) \ge (2R+1)/(6R+7).$$

Bounds for $\zeta(5/7 + it)$. The technique of the last section can be adapted to bound $\zeta(\sigma + it)$ for any fixed σ in the range $1/2 < \sigma < 1$. To illustrate this, we will show that

$$\zeta(5/7 + it) \ll t^{(7R-3)/(14R+28)+\epsilon} \tag{5.4.2}$$

for every $\epsilon > 0$. For comparison purposes, note that

$$\frac{7R-3}{14R+28} = .07077534556548856\ldots$$

In Theorem 2.12, we obtained the exponent

$$\frac{1}{14} = .07142857142857143\ldots$$

and in Theorem 4.2 we obtained the exponent

$$\frac{368}{5159} = .07133165342120566\ldots.$$

If (k,l) is an exponent pair and $N < N_1 \le 2N$ then

$$\sum_{N<n\le N_1} n^{-5/7-it} \ll t^{\theta_\alpha(k,l)},$$

where

$$\theta_\alpha(k,l) = \frac{(1-\alpha)k + \alpha l - 5\alpha}{7}.$$

Let

$$g(\alpha) = \inf \theta_\alpha(k,l),$$

and set $\delta = \sup\{g(\alpha) : 0 \le \alpha \le 1\}$. We will show that $\delta = (7R-3)/(14R+28)$. By Lemma 2.11, this is sufficient to prove (5.4.2).

Suppose first that $\alpha \le 1/3$, and let

$$(k,l) = A^2BA^2BAB(0,1) = (\frac{11}{186}, \frac{25}{31}).$$

Since $l - k - 5/7 > 0$, it follows that $\theta_\alpha(k,l)$ is an increasing function of α, and so

$$g(\alpha) \le \theta_{\frac{1}{3}}(k,l) = \frac{137}{1953} < \frac{7R-3}{14R+28}$$

for $\alpha \le 1/3$.

Next, suppose that $\alpha \ge 9/25$, and let

$$(k,l) = ABABA^2BAB(0,1) = (\frac{4}{49}, \frac{75}{98}).$$

Since $l - k + 5/7 < 0$, we see that

$$g(\alpha) \le \theta_{\frac{9}{25}}(k,l) = \frac{173}{2450} < \frac{7R-3}{14R+28}$$

for $\alpha \ge 9/25$.

Henceforth, we assume that $1/3 < \alpha < 9/25$. For α in this range, we have

$$\inf \theta_\alpha = \inf \theta_\alpha A = \frac{7k + (7l - 10k - 3)\alpha}{14k + 14}.$$

Let $\psi_\alpha(k,l) = \theta_\alpha A(k,l)$. We claim that for every $(k,l) \in \mathbf{AP}$, $\psi_\alpha(k,l)$ is an increasing function of α, and that for every $(k,l) \in B\mathbf{AP}$, $\psi_\alpha(k,l)$ is a decreasing function of α. For if $(k,l) = A(\kappa,\lambda)$, then

$$7l - 10k - 3 = \frac{7(\lambda-\kappa) + (1-2\kappa)}{2\kappa+2} \ge 0, \tag{5.4.3}$$

and if $(k,l) = BA(\kappa,\lambda)$ then

$$7l - 10k - 3 = \frac{16(\kappa-\lambda)+(2-4\lambda)}{2\kappa+2} \leq 0. \tag{5.4.4}$$

We will now show that

$$\inf \psi_\alpha \leq \frac{7R-3}{14R+28}$$

for all α under consideration. Let $(k_1,l_1) \in A\mathbf{P}$ and let $(k_2,l_2) \in BA\mathbf{P}$. We can find a convex combination (k,l) of (k_1,l_1) and (k_2,l_2) such that $7l-10k-3=0$. For this exponent pair,

$$\psi_\alpha(k,l) = \frac{k}{2k+2} = \frac{7(k+l)-3}{14(k+l)+28}.$$

Since

$$\psi_\alpha(k,l) \geq \min(\psi_\alpha(k_1,l_1), \psi_\alpha(k_2,l_2)),$$

it follows that

$$\inf \psi_\alpha \leq \inf \frac{7(k+l)-3}{14(k+l)+28} = \frac{7R-3}{14R+28}.$$

To complete the proof, we will now show that there is some β with $1/3 < \beta < 9/25$ and

$$\inf \psi_\beta \geq \frac{7R-3}{14R+28}. \tag{5.4.5}$$

Now let $g_A(\alpha) = \inf \psi_\alpha A$ and $g_B(\alpha) = \inf \psi_\alpha BA$, so that $\inf \psi_\alpha = \min(g_A(\alpha), g_B(\alpha))$. From the inequalities (5.4.3) and (5.4.4), we see that g_A is an increasing function of α and that g_B is a decreasing function of α. From Theorem 5.5, we see that $g_A(1/3) < g_B(1/3)$ and $g_B(9/25) < g_A(9/25)$. Since g_A and g_B are continuous, there is some β with $1/3 < \beta < 9/25$ and $g_A(\beta) = g_B(\beta)$.

Let (k'_n, l'_n) be a sequence of exponent pairs in $A\mathbf{P}$ such that

$$\inf \psi_\beta A = \lim_{n\to\infty} (k'_n, l'_n).$$

Let (k''_n, l''_n) be a sequence of exponent pairs in $BA\mathbf{P}$ such that

$$\inf \psi_\beta BA = \lim_{n\to\infty} (k''_n, l''_n).$$

Since $7l'_n - 10k'_n - 3 \geq 0$ and $7l''_n - 10k''_n - 3 \leq 0$, we can find a convex combination (k_n, l_n) of (k'_n, l'_n) and (k''_n, l''_n) such that $7l_n - 10k_n - 3 = 0$. We then have

$$\psi_\beta(k_n,l_n) = \frac{7((k_n+l_n)-3}{14((k_n+l_n)+28}.$$

Since $\psi_\beta(k_n,l_n) \leq \max(\psi_\beta(k'_n,l'_n), \psi_\beta(k''_n,l''_n))$, statement (5.4.5) follows.

5.5 NOTES

The material in this chapter is taken from Graham (1985), which was the first published account of computing optimal exponent pairs for general θ. Phillips and Rankin had earlier considered the special case $\theta(k,l) = k + l$. Phillips (1935) stated that he computed the first 12 terms of the q-sequence. He did not publish them; he stated only that he failed to find any pattern. Rankin (1955) gave the first eleven terms of the q-sequence for $\inf(k+l)$, and this gives the numerical value to 10 decimal places. He did the calculations during the 1940's on an adding machine. He did not publish the details of his method; the proofs, he said, "are rather long and involve much heavy algebra." He also stated that all the terms in the q-sequence of $\inf(k+l)$ satisfy $q \leq 3$. Heath-Brown (private communication) has established a corresponding result for more general θ: if θ satisfies (5.2.2), then every term in the q-sequence after the first is ≤ 3.

6. TWO DIMENSIONAL EXPONENTIAL SUMS

6.1 INTRODUCTION

In this section, we shall give a brief overview of the material in this chapter.

Let $\mathbf{D}$ be a subset of $[X, 2X] \times [Y, 2Y]$. In most applications, $\mathbf{D}$ is a rectangle, or some set bounded by simple algebraic curves. Let $f : \mathbf{D} \to \mathbf{R}$ and define

$$S = \sum_{(m,n) \in \mathbf{D}} e(f(m,n)).$$

In analogy with our hypotheses for exponent pairs in the one dimensional case, it is appropriate to assume that

$$D_{x^i y^j} f(x,y) = D_{x^i y^j}(Ax^{-\alpha} y^{-\beta}) \cdot \{1 + O(\Delta)\} \tag{6.1.1}$$

for all i, j less than some appropriate bound, and where

$$D_{x^i y^j} = \frac{\partial^{i+j}}{\partial x^i \partial y^j},$$

A is a non-zero real constant, $\alpha < 1$, $\beta < 1$, $\alpha\beta \neq 0$, and $\Delta = \Delta(X, Y) \to 0$ as $X \to \infty$ and $Y \to \infty$. The primary tools for estimating S are two dimensional analogues of the Poisson summation formula and the Weyl-van der Corput inequality.

First, let us consider the Poisson summation formula. Recall that in Lemma Lemma 3.6, terms of the form $|f''(x_\nu)|^{-1/2}$ appear, so the usefulness of that lemma is lessened when f'' becomes small. In two dimensional sums, the Hessian of f plays a similar role. The Hessian of f is defined by

$$Hf = \det \begin{bmatrix} D_{xx} f & D_{xy} f \\ D_{xy} f & D_{yy} f \end{bmatrix}.$$

Under suitable conditions on f and $\mathbf{D}$, we have

$$\sum_{(m,n) \in \mathbf{D}} e(f(m,n)) \ll M^{-1/2} \Big| \sum_{(\mu,\nu) \in \mathbf{D}''} e(f(\xi, \eta) - \mu\xi - \nu\eta) \Big| + \text{error terms},$$

where

(i) $Hf \approx M$,

(ii) $\mathbf{D}'$ is the image of $\mathbf{D}$ under $\mu = D_x f$ and $\nu = D_y f$,

(iii) $\mathbf{D}''$ is some subset of $\mathbf{D}$,

(iv) $\xi = \xi(\mu,\nu)$ and $\eta = \eta(\mu,\nu)$ are defined by $D_x f(\xi,\eta) = \mu$ and $D_y f(\xi,\eta) = \nu$.

The two dimensional Weyl-van der Corput inequality states that if $Q \le X$ and $R \le Y$ then

$$|S|^2 \ll \frac{X^2Y^2}{QR} + \frac{XY}{QR} \sum_{|q|<Q} \sum_{\substack{|r|<R \\ (q,r)\neq(0,0)}} |S_1(q,r)|, \tag{6.1.2}$$

where

$$S_1(q,r) = \sum_{(m,n)\in \mathbf{D}_1(q,r)} e(f_1(m,n;q,r)),$$

$$f_1(m,n;q,r) = f(m+q,n+r) - f(m,n),$$

and

$$\mathbf{D}_1(q,r) = \{(m,n) : (m+qt, n+rt) \in \mathbf{D} \text{ for } t=0 \text{ and } t=1\}.$$

In analogy with the one dimensional case, we can hope to prove an estimate of the form

$$S \ll L_1^{k_1} X^{l_1} L_2^{k_2} Y^{l_2}, \tag{6.1.3}$$

where

$$L_1 = |A| X^{-\alpha-1} Y^{-\beta} \text{ and } L_2 = |A| X^{-\alpha} Y^{-\beta-1} .$$

Note that $L_1 \approx D_x f$ and $L_2 \approx D_y f$. If we can prove an estimate of the form (6.1.3) under appropriate assumptions on f and $\mathbf{D}$, we say that $(k_1,l_1;k_2,l_2)$ is an *exponent quadruple.* Note that since

$$|S| \ll \sum_m \Big|\sum_n e(f(m,n))\Big|, \tag{6.1.4}$$

$(0,1;k,l)$ is an exponent quadruple whenever (k,l) is an exponent pair. Similarly, $(k,l;0,1)$ is an exponent quadruple.

Unfortunately, the application of the Weyl-van der Corput inequality and the Poisson summation formula is not as straightforward as it is in the one dimensional case. To illustrate why this is so, we consider

$$f(m,n) = Am^{-\alpha}n^{-\beta}.$$

After applying Weyl-van der Corput, we encounter functions of the form

$$f_1(m,n;q,r) = \int_0^1 \frac{d}{dt} A(m+qt)^{-\alpha}(n+rt)^{-\beta} dt \sim -Am^{-\alpha}n^{\beta}\left(\frac{\alpha q}{m} + \frac{\beta r}{n}\right).$$

If we then apply the Poisson summation formula, we must first compute Hf_1. Now

$$Hf_1 \sim A^2 m^{-2\alpha} n^{-2\beta} \alpha\beta(\alpha+\beta+2)P$$

where

$$P = \frac{(\alpha)_2 q^2}{m^2} + \frac{2(\alpha+1)(\beta+1)qr}{mn} + \frac{(\beta)_2 r^2}{n^2}.$$

For some values of the parameters, P will vanish or be inconveniently small. The effect of this is that the Poisson summation formula cannot be applied directly. Instead, we subdivide $\mathbf{D}$ into a region where P is small and another region where P is large. In the latter region, we can apply the Poisson summation formula. In the former region, we use some othér estimate such as (6.1.4). There are technical difficulties in carrying this out, and the difficulties become more pronounced the more often that Weyl-van der Corput is used.

Here we shall ignore these difficulties and argue heuristically. The function f_1 satisfies

$$f_1(m,n;q,r) = \int_0^1 \frac{d}{dt} f(m+qt, n+rt)dt \approx \frac{qF}{X} + \frac{rF}{Y} \approx \rho F$$

where $F = |A| X^{-\alpha} Y^{-\beta}$ and $\rho = \max(|q|/X, |r|/Y)$. If $(k_1, l_1; k_2, l_2)$ is an exponent quadruple, then

$$S_1(q,r) \ll (F\rho)^{k_1+k_2} X^{l_1-k_1} Y^{l_2-k_2}.$$

Now assume that Q and R are chosen so that $Q/X = R/Y$. If we set $Z = Q^2Y/X = R^2X/Y$, then

$$\frac{1}{QR} \sum_{|q|<Q} \sum_{|r|<R} \rho^{k_1+k_2} \ll \left(\frac{Z}{XY}\right)^{(k_1+k_2)/2}.$$

We plug this back into (6.1.2) to obtain

$$S^4 \ll X^4Y^4Z^{-2} + F^{2k_1+2k_2} X^{2+2l_1-3k_1-k_2} Y^{2+2l_2-k_1-3k_2} Z^{k_1+k_2}.$$

Choose Z so that the two terms on the right hand side are equal. Then

$$S^{2k_1+2k_2+4} \ll (F/X)^{2k_1} X^{k_1+k_2+2l_1+2} (F/Y)^{2k_2} Y^{k_1+k_2+2l_2+2}.$$

Thus we see heuristically that if $(k_1, l_1; k_2, l_2)$ is an exponent quadruple then so is

$$A(k_1, l_1; k_2, l_2) = \frac{(2k_1, k_1+k_2+2l_1+2; 2k_2, k_1+k_2+2l_2+2)}{2k_1+2k_2+4}.$$

Similarly, a heuristic argument with the Poisson summation formula yields the exponent quadruple

$$B(k_1, l_1; k_2, l_2) = (l_1 - 1/2, k_1 + 1/2; l_2 - 1/2, k_2 + 1/2).$$

One way of avoiding the difficulties implicit in the application of Weyl-van der Corput is to apply it with $Q = 1$ or $R = 1$. By taking $R = 1$ and arguing heuristically, we see that this approach should lead to the exponent quadruple

$$A_1(k_1, l_1; k_2, l_2) = \frac{(k_1, k_1 + k_2 + l_1 + 1; k_2, 2k_1 + 2k_2 + l_2 + 1)}{2k_1 + 2k_2 + 2}.$$

Similarly, with $Q = 1$ one gets

$$A_2(k_1, l_1; k_2, l_2) = \frac{(k_1, 2k_1 + 2k_2 + l_1 + 1; k_2, k_1 + k_2 + l_2 + 1)}{2k_1 + 2k_2 + 2}.$$

We may use convexity and take the average of A_1 and A_2 to get the exponent quadruple

$$A_S(k_1, l_1; k_2, l_2) = \frac{(k_1, k_1 + k_2 + l_1 + 1; k_2, 2k_1 + 2k_2 + l_2 + 1)}{2k_1 + 2k_2 + 2}.$$

The "S" here stands for Srinivasan, who used essentially this operation in method of exponent pairs.

It is also possible to apply the Poisson summation formula to one variable at a time and get the exponent quadruples

$$B_1(k_1, l_1; k_2, l_2) = (l_1 - 1/2, k_1 + 1/2; k_2, l_2)$$

and

$$B_2(k_1, l_1; k_2, l_2) = (k_1, l_1; l_2 - 1/2, k_2 + 1/2).$$

In some applications, the critical cases for estimating S occur when $X \approx Y$. In such cases, it is desirable to have $k_1 = k_2$ and $l_1 = l_2$. Note that

$$A(k, l; k, l) = (\frac{k}{2k + 2}, \frac{k + l + 1}{2k + 2}; \frac{k}{2k + 2}, \frac{k + l + 1}{2k + 2})$$

and

$$B(k, l; k, l) = (l - 1/2, k + 1/2; l - 1/2, k + 1/2).$$

We therefore have the following

CONJECTURE Q. *If (k, l) is an exponent pair, then $(k, l; k, l)$ is an exponent quadruple.*

In this chapter, we will prove Conjecture Q in the special cases $(k, l) = B(0, 1) = (1/2, 1/2)$ and $(k, l) = AB(0, 1) = (1/6, 2/3)$.

Srinivasan (1965) used A_S to develop a theory of exponent quadruples. Roughly stated, his theory is as follows. Let $\mathbf{P_S}$ be the set of all pairs obtained from $(0, 1)$ by

$$A_S(k, l) = (\frac{k}{4k + 2}, \frac{3k + l + 1}{4k + 2})$$

and

$$B(k,l) = (l - 1/2, k + 1/2).$$

If $(k,l) \in \mathbf{P_S}$, then $(k,l;k,l)$ is an exponent quadruple.

The applications of Chapter 4 can be done with two dimensional sums. Assume that $(k,l;k,l)$ is an exponent quadruple, and let

$$\theta = \theta(k,l) = \frac{2k + 2l - 1}{4l - 1}.$$

Then for some constant $C > 0$, we have

$$\zeta(1/2 + it) \ll t^{\theta/2} \log^C t,$$

$$\Delta(x) \ll x^{\theta} \log^C x + x^{1/4} \log x,$$

$$R(x) \ll x^{\theta} \log^C x + x^{1/4} \log x.$$

Here is a historical survey of the results of this type that have appeared in the literature.

1. $(k,l) = A_S^3 B(0,1)$; $\theta = 19/58$. This was done by Titchmarsh (1942) for $\zeta(1/2 + it)$.
2. $(k,l) = A_S^2 AB(0,1)$; $\theta = 15/46$. This was done by Titchmarsh (1934) for $R(x)$, by Min (1949) for $\zeta(1/2 + it)$, and by Richert (1953) for $\Delta(x)$.
3. $(k,l) = A_S A^2 B(0,1)$; $\theta = 13/40$. This was done by Hua (1941) for $R(x)$.
4. $(k,l) = A^3 B(0,1)$; $\theta = 12/37$. This was done by Haneke (1963) for $\zeta(1/2 + it)$, by Chen (1963) for $R(x)$, and by Kolesnik (1969) for $\Delta(x)$.
5. $(k,l) = A^3 A_S^3 B(0,1)$; $\theta = 35/108$. This was done by Kolesnik (1982) for $\zeta(1/2 + it)$ and $\Delta(x)$.

Note that $35/108 = .324\overline{074}$. If we assume Conjecture Q for all (k,l) and apply the algorithm of Chapter 5, we find that the optimal q-sequence is

$$32122\ 11121\ 21211\ 11122\ 11111\ldots,$$

the limiting value for θ is

$$0.32392\ 45703\ 76329\ 83494\ 00175\ 84916\ldots,$$

and the limiting value for $\theta/2$ is

$$0.16196\ 22851\ 88164\ 491747\ 00087\ 92458\ldots.$$

Bombieri and Iwaniec (1986a,b) proved that

$$\zeta(1/2 + it) \ll t^{9/56+\epsilon}. \tag{6.1.5}$$

Since $9/56 = .160\overline{7142857}$, their method is superior to the two dimensional method. We shall give a proof of (6.1.5) in Chapter 7. Moreover, Iwaniec and Mozzochi (1988) have proved that

$$\Delta(x) \ll x^{7/22+\epsilon}$$

and

$$R(x) \ll x^{7/22+\epsilon}.$$

Since $7/22 = .3\overline{18}$, this supersedes two dimensional methods even further.

6.2 GENERALIZED WEYL-VAN DER CORPUT INEQUALITY

Our first lemma is a generalized version of the Weyl-van der Corput inequality. We will be using this lemma only in $\mathbf{R}^2$, but we prove it for domains in $\mathbf{R}^k$, since there is no extra effort involved in generalizing it to k dimensions.

LEMMA 6.1. *Suppose that* $\mathbf{D}$ *and* $\mathbf{Q}$ *are subsets of* $\mathbf{R}^k$. *Assume further that* $\xi(\mathbf{n})$ *is a function supported on* $\mathbf{D}$ *and* $\eta(\mathbf{q})$ *is a function supported on* $\mathbf{Q}$ *such that* $\sum_{\mathbf{q}} \eta(\mathbf{q}) = 1$. *Define*

$$S = \sum_{\mathbf{n}} \xi(\mathbf{n}),$$

$$\mathbf{D}' = \{\mathbf{n} \in \mathbf{R}^k : \mathbf{n} + \mathbf{q} \in \mathbf{D} \text{ for some } \mathbf{q} \in \mathbf{Q}\},$$

$$\mathbf{Q}' = \{\mathbf{h} = \mathbf{q} - \mathbf{r} : \mathbf{q}, \mathbf{r} \in \mathbf{Q}\},$$

and

$$a(\mathbf{h}) = \sum_{\mathbf{q}-\mathbf{r}=\mathbf{h}} \eta(\mathbf{q})\overline{\eta(\mathbf{r})}.$$

Then

$$|S|^2 \le |\mathbf{D}'| \sum_{\mathbf{h}\in\mathbf{Q}'} a(\mathbf{h}) \sum_{\mathbf{n}} \xi(\mathbf{n}+\mathbf{h})\overline{\xi(\mathbf{n})}.$$

Proof. From the hypothesis that $\sum_{\mathbf{q}} \eta(\mathbf{q}) = 1$, we see that

$$S = \sum_{\mathbf{n}} \xi(\mathbf{n}) \sum_{\mathbf{q}} \eta(\mathbf{q}) = \sum_{\mathbf{q}} \sum_{\mathbf{n}} \eta(\mathbf{q})\xi(\mathbf{n}+\mathbf{q}) = \sum_{\mathbf{n}\in\mathbf{D}'} \sum_{\mathbf{q}\in\mathbf{Q}} \eta(\mathbf{q})\xi(\mathbf{n}+\mathbf{q}).$$

By Cauchy's inequality,

$$\begin{aligned} |S|^2 &\le |\mathbf{D}'| \sum_{\mathbf{n}\in\mathbf{D}'} \Big|\sum_{\mathbf{q}\in\mathbf{Q}} \eta(\mathbf{q})\xi(\mathbf{n}+\mathbf{q})\Big|^2 \\ &= |\mathbf{D}'| \sum_{\mathbf{q},\mathbf{r}\in\mathbf{Q}} \sum_{\mathbf{n}} \xi(\mathbf{n}+\mathbf{q})\overline{\xi(\mathbf{n}+\mathbf{r})} \end{aligned}$$

Now let $\mathbf{m} = \mathbf{n} + \mathbf{r}$ and $\mathbf{h} = \mathbf{q} - \mathbf{r}$ to get

$$|S|^2 \le |\mathbf{D}'| \sum_{\mathbf{h}\in\mathbf{Q}'} a(\mathbf{h}) \sum_{\mathbf{m}} \xi(\mathbf{m}+\mathbf{h})\overline{\xi(\mathbf{m})}$$

as asserted.

If we take $k = 2$ in the above lemma, $\mathbf{Q} = [1, Q] \times [1, R]$, and $\eta(q,r) = 1/(QR)$, then we obtain

$$|S|^2 \ll \frac{|\mathbf{D}'|}{QR} \sum_{|q| \le Q} \sum_{|r| \le R} \Big| \sum_{(m,n) \in \mathbf{D}(q,r)} e(f_1(m,n;q,r)) \Big|,$$

where

$$\mathbf{D}' = \{(m,n) : (m+q, n+r) \in \mathbf{D} \text{ for some } (q,r) \text{ with } |q| \le Q, |r| \le R\}$$

$$f_1(m,n;q,r) = \int_0^1 \frac{\partial}{\partial t} f(m+tq, n+tr)\, dt,$$

and

$$\mathbf{D}(q,r) = \{(m,n) : (m+tq, n+tr) \in \mathbf{D} \text{ for } t = 0 \text{ and } t = 1\}.$$

In the next four lemmas, we shall obtain estimates for $\mathbf{D}'$ for the kinds of sets that we shall be working with.

LEMMA 6.2. *Suppose*

$$\mathbf{D} \subseteq \{(m,n) : X < m \le X + M\,, Y < n \le Y + N\},$$

where $M \le X$ and $N \le Y$. Let Z be a parameter satisfying

$$Z \le \frac{M^2 N^2}{XY}.$$

Define

$$Q = \sqrt{ZX/Y} \text{ and } R = \sqrt{ZY/X}.$$

Then $|\mathbf{D}'| \ll MN$.

Proof. If $(m,n) \in \mathbf{D}'$ then there exists q and r such that $X < m + q \le X + M$, $Y < n + r \le Y + N$, $1 \le q \le Q$, and $1 \le r \le R$. Therefore

$$X - Q < m \le X + M - 1 \text{ and } Y - R < n \le Y + N - 1.$$

Consequently $|bD'| \le (M+Q)(N+R)$. Since $Q \le M\sqrt{N/Y} \le M$ and $R \le N\sqrt{M/X} \le N$, we see that $|\mathbf{D}'| \le 4MN$.

LEMMA 6.3. *Let a and b be non-zero complex numbers, and let δ be a positive real number. Define*

$$\mathbf{E} = \mathbf{E}(\delta; a, b) = \{(m,n) : X < m \le 2X, Y < n \le 2Y, \text{ and } |an - bm| < \delta m\}.$$

Then

$$|\mathbf{E}| \ll |b|^{-1}\delta XY + 1.$$

Proof. First observe that $\mathbf{E}(\delta/|b|; a/b, 1) = \mathbf{E}(\delta; a, b)$. Thus it suffices to consider the case $b = 1$. We may also assume that $\delta \ll 1$, that $\mathbf{E} \neq \emptyset$, and that $|a| \approx X/Y$; otherwise, the estimate is trivial.

For any $(m, n) \in \mathbf{E}$,

$$\left|\frac{n}{m} - \frac{1}{a}\right| \leq \frac{\delta}{|a|}.$$

Suppose $(m_0, n_0) \in \mathbf{E}$,and let $k = \gcd(m_0, n_0)$. Define $m_1 = m_0/k$ and $n_1 = n_0/k$. Then $\gcd(m_1, n_1) = 1$, and for any other $(m, n) \in \mathbf{E}$, we must have

$$\left|\frac{n}{m} - \frac{n_1}{m_1}\right| \leq \frac{2\delta}{|a|}$$

and

$$|nm_1 - mn_1| \leq \frac{8\delta X^2}{|a|k}.$$

For every l, there is exactly one solution (m, n) of the equation $nm_1 - n_1m = l$ with $0 < m \leq m_1$ and $0 < n \leq n_1$, and $\ll k$ solutions of $nm_1 - n_1m = l$ with $m \approx X$ and $n \approx Y$. Thus there are

$$\ll \frac{\delta X^2}{|a|k} \cdot k \approx \delta XY$$

pairs $(m, n) \in \mathbf{E}$ different from (m_0, n_0). This proves the lemma.

LEMMA 6.4. *Let $P(x) = a_k x^k + \ldots + a_1 x + a_0$ be a polynomial of degree k with complex coefficients. Let $\zeta_1, \ldots, \zeta_k$ be the zeros of P. Define*

$$B = \min_{1 \leq i \leq k} |\zeta_i P'(\zeta_i)|.$$

Let q and r be non-zero integers. For $\delta > 0$, define

$$\mathbf{E} = \left\{(m, n) : X < m \leq 2X, Y < n \leq 2Y, \left|P\left(\frac{qn}{rm}\right)\right| < \delta\right\}.$$

Then

$$|\mathbf{E}| \ll k2^k B^{-1}\delta XY + k,$$

and the implied constant is absolute.

Note that the stated bound is trivial if $P(0) = 0$ or if P has a multiple zero.

Proof. For $j = 1, \ldots, k$ define

$$\mathbf{E}_j = \left\{(m, n) \in \mathbf{E} : \left|\frac{qn}{rm} - \zeta_j\right| \leq \left|\frac{qn}{rm} - \zeta_i\right| \text{ for } 1 \leq i \leq k\right\}.$$

In other words, $\mathbf{E}_j$ is the set of (m,n) in $\mathbf{E}$ such that ζ_j is the closest root to qn/rm. If $(m,n) \in \mathbf{E}_j$ then

$$|\zeta_i - \zeta_j| - \left|\frac{qn}{rm} - \zeta_i\right| \leq \left|\frac{qn}{rm} - \zeta_j\right| \leq \left|\frac{qn}{rm} - \zeta_i\right|$$

by the triangle inequality. Consequently

$$\left|\frac{qn}{rm} - \zeta_i\right| \geq \frac{1}{2}|\zeta_i - \zeta_j|$$

for $i \neq j$. Moreover

$$\begin{aligned}\delta \geq \left|P(\frac{qn}{rm})\right| &= \left|a_k(\frac{qn}{rm} - \zeta_1)\dots(\frac{qn}{rm} - \zeta_k)\right| \\ &\geq 2^{1-k}|a_k|\left|\frac{qn}{rm} - \zeta_j\right| \prod_{\substack{i=1 \\ i\neq j}}^{k} |\zeta_i - \zeta_j| = 2^{1-k}|P'(\zeta_j)|\left|\frac{qn}{rm} - \zeta_j\right|.\end{aligned}$$

Now set

$$\mathbf{E}'_j = \left\{(m,n) : X < m \leq 2X, Y < n \leq 2Y, \text{ and } |\frac{qn}{r} - \zeta_j m| \leq 2^{k-1}\delta|P'(\zeta_j)|^{-1}m\right\}. \tag{6.2.1}$$

Then $\mathbf{E} \subseteq \bigcup_{j=1}^{k} \mathbf{E}'_j$. By Lemma 6.3, $|\mathbf{E}'_j| \ll 2^k B^{-1}\delta XY + 1$, so $|\mathbf{E}| \ll k2^k B^{-1}XY + k$.

LEMMA 6.5. *Let $P(x) = a_k x^k + \dots + a_0$ and $Q(x) = b_l x^l + \dots + b_0$ be polynomials with complex coefficients. Let $\zeta_1, \dots, \zeta_k$ be the zeros of P and let $\eta_1, \dots, \eta_l$ be the zeros of Q. Define*

$$C_1 = \{|P'(\zeta_i)| : 1 \leq i \leq k\}, C_2 = \{|Q'(\eta_i)| : 1 \leq j \leq l\}, \text{ and } C = \min C_1 \cup C_2.$$

Let q and r be integers with $r \neq 0$, and set

$$\mathbf{E} = \left\{(m,n) : X < m \leq 2X, Y < n \leq 2Y, \text{ and } |P(\frac{qn}{rm})| \leq \delta\right\},$$
$$\mathbf{F} = \left\{(m,n) : X < m \leq 2X, Y < n \leq 2Y, \text{ and } |Q(\frac{qn}{rm})| \leq \delta\right\}.$$

If $|\zeta_i - \eta_j| \geq 2^{\max(k,l)}C^{-1}\delta$ for all i and j, then $\mathbf{E} \cap \mathbf{F} = \emptyset$.

Proof. Without loss of generality, we may assume that $l \geq k$. Let $\mathbf{E}'_j$ be as defined in (6.2.1), and let $\mathbf{F}'_i$ be defined analogously. From the proof of Lemma 6.4, we see that $\mathbf{E} \subseteq \bigcup_{j=1}^{k} \mathbf{E}'_j$ and $\mathbf{F} \subseteq \bigcup_{i=1}^{l} \mathbf{F}'_i$. By hypothesis, $\mathbf{E}'_i \cap \mathbf{F}'_j = \emptyset$ for all i and j, so $\mathbf{E} \cap \mathbf{F} = \emptyset$.

6.3 OMEGA CONDITIONS

Assume that $f(u,v)$ is defined and real-valued on the rectangle $[X,2X]\times[Y,2Y]$ and $\mathbf{D}$ is a subdomain of this rectangle. We wish to estimate

$$S=\sum_{(m,n)\in\mathbf{D}} e(f(m,n)),$$

and we shall need several assumptions on $\mathbf{D}$ and f. For convenience, we list these assumptions as $(\Omega_1),(\Omega_2)$, and (Ω_3). We begin by listing the first two conditions.

(Ω_1) f has partial derivatives of all orders, and there is some constant C_1 such that

$$\left|\frac{\partial^{a+b}f}{\partial u^a\partial v^b}\right|\le C_1FX^{-a}Y^{-b}$$

for all a and b with $0\le a\le 4$ and $0\le b\le 4$.

(Ω_2) There is some constant C_2 such that for each fixed n, $Y<n\le 2Y$, the set

$$I(n)=\{m:(m,n)\in\mathbf{D}\}$$

consists of at most C_2 intervals. Similarly, for each fixed m, the set

$$J(m)=\{n:(m,n)\in\mathbf{D}\}$$

consists of at most C_2 intervals.

When using the Ω-conditions, we will allow the implied constants to depend on the C_i's, but all estimates will be uniform in F,X, and Y.

LEMMA 6.6. *Assume (Ω_1) and (Ω_2). Assume also that $f_{xx}\approx\Lambda$ on $\mathbf{D}$. Then*

$$S\ll|\mathbf{D}|\Lambda^{1/2}+\Lambda^{-1/2}Y.$$

Proof. Write

$$S=\sum_{Y<n\le 2Y}\sum_{m\in I(n)} e(f(m,n))$$

and use Theorem 2.2 on the inner sum to obtain

$$S\ll\sum_{Y<n\le 2Y}\sum_{m\in I(n)}(|I(n)|\Lambda^{1/2}+\Lambda^{-1/2})\ll|\mathbf{D}|\Lambda^{1/2}+Y\Lambda^{-1/2}.$$

We define $\psi(x,y)$ by the relation

$$f_x(\psi(x,y),y)=x, \tag{6.3.1}$$

and we will assume that

(Ω_3) There is a constant C_3 such that for each fixed l, the set

$$J^*(l) = \{n : (\psi(l,n), n) \in \mathbf{D}\}$$

consists of at most C_3 intervals.

Let

$$g(w,v) = f(\psi((w,v)), v) - w\psi((w,v), \qquad (6.3.2)$$

and let Hf denote the Hessian of f; i.e.

$$Hf = f_{xx}f_{yy} - f_{xy}^2.$$

Our next lemma expresses the first and second derivatives of g in terms of f.

LEMMA 6.7. *Let g be as defined above. Then*

$$g_y = f_y(\psi, y) \text{ and } g_{yy} = \frac{Hf(\psi, y)}{f_{xx}(\psi, y)}.$$

Proof. We differentiate (6.3.1) with respect to y to get

$$f_{xx}(\psi, y)\psi_y + f_{xy}(\psi, y) = 0,$$

whence

$$\psi_y = \frac{-f_{xy}(\psi, y)}{f_{xx}(\psi, y)}. \qquad (6.3.3)$$

Differentiation of g with respect to v yields

$$g_y = f_x\psi_y + f_y - f_x\psi_v = f_y.$$

This proves the first assertion. We differentiate again to get $g_{yy} = f_{xy}\psi_y + f_{yy}$, and the second assertion follows by combining this with (6.3.3).

LEMMA 6.8. *Assume* $(\Omega_1), (\Omega_2)$, and (Ω_3). *Let L denote* $\log(FXY + 2)$. *Assume also that*

$$f_{xx} \approx \Lambda, Hf \approx M, \text{ and } |f_u| \geq 1/2$$

on $\mathbf{D}$. *Then*

$$S \ll |\mathbf{D}|M^{1/2} + FM^{-1/2}X^{-1} + Y\Lambda^{-1/2}L + YL.$$

Proof. Lemma 6.6 proves the estimate when $\Lambda \leq M$. Henceforth we assume

$$\Lambda \geq M. \qquad (6.3.4)$$

We also assume that $f_{xx} < 0$ on $\mathbf{D}$. We can do this by replacing f by $-f$ if necessary.

From Lemma 3.6, we see that

$$S = \sum_{Y<n\le 2Y}\ \sum_{m\in I(n)} e(f(m,n))$$
$$= \sum_{Y<n\le 2Y}\ \sum_{k\in K(n)} e(-1/8+g(k,n))|f_{xx}(\psi,n)|^{-1/2} + O(Y\Lambda^{-1/2}+YL),$$

where $K(n)=\{k:(\psi(k,n),n)\in \mathbf{D}\}$. Changing the order of summation yields

$$S = \sum_{K_1\le k\le K_1+K}\ \sum_{n\in J^*(k)} e(-1/8+g(k,n))|f_{xx}|^{-1/2} + O(Y\Lambda^{-1/2}+YL)$$

for some appropriate K and K_1. From (Ω_1) with $a=0$ and $b=1$, we see that $K \ll FX^{-1}$ and $K_1 \ll FX^{-1}$. Using partial summation and Theorem 2.2 for the inner sum, we get

$$S \ll \sum_{K_1\le k\le K_1+K} \Lambda^{-1/2}(|J^*(k)|M^{1/2}\Lambda^{-1/2}+M^{-1/2}\Lambda^{1/2}) + Y\Lambda^{-1/2}+YL. \quad (6.3.5)$$

Furthermore,

$$\sum_k |J^*(k)| = \sum_n \sum_{\substack{k \\ \psi(k,n)\in I(n)}} 1.$$

Now the inner sum is the number of values assumed by the function $f_x(m,n)$ as m runs through the interval $I(n)$. Therefore this inner sum is

$$\ll \int_{I(n)} f_{uu}(t,n)dt + 1 \ll |I(n)|\Lambda + 1,$$

and so

$$\sum_k |J^*(k)| \ll \sum_{Y<n\le 2Y} (|I(n)|\Lambda+1) \ll |\mathbf{D}|\Lambda + Y.$$

We combine this with (6.3.5) to get

$$S \ll |\mathbf{D}|M^{1/2} + \Lambda^{-1}M^{1/2}Y + FM^{-1/2}X^{-1} + Y\Lambda^{-1/2} + YL.$$

By (6.3.4), $Y\Lambda^{-1/2} \ge \Lambda^{-1}M^{1/2}Y$, so the desired estimate follows.

LEMMA 6.9. *Suppose that there is some constant c such that $|f_x| \le cFX^{-1} \le 1/2$ on $\mathbf{D}$. Then*

$$S \ll F^{-1}XY.$$

Proof. We again note that

$$S = \sum_{Y<n\le 2Y}\ \sum_{m\in I(n)} e(f(m,n)).$$

By the Kusmin-Landau inequality (Theorem 2.1), the inner sum is $\ll F^{-1}X$.

LEMMA 6.10. *Assume* $(\Omega_1), (\Omega_2)$, and (Ω_3). *Assume also that*

$$f_x \approx FX^{-1}, f_{xx} \approx FX^{-2}, \text{ and } Hf \approx \delta F^2 X^{-2} Y^{-2}$$

for some $\delta \leq 1$. *Let* $XY = N$. *Then*

$$S \ll |\mathbf{D}|\delta^{1/2} F N^{-1} + F^{-1/2} NL + \delta^{-1/2} YL.$$

Proof. If $F < 1$ then the trivial estimate gives

$$S \ll |\mathbf{D}| \ll N \ll F^{-1/2} NL.$$

If $1 \leq F \leq cX$ and c is sufficiently small, then Lemma 6.9 may be applied to yield

$$S \ll F^{-1} XY \ll F^{-1/2} NL.$$

Assume now that $F > cX$. We can then apply Lemma 6.8 with $\Lambda = FX^{-2}$ and $M = \delta^2 F^2 N^{-2}$ to get

$$S \ll |\mathbf{D}|\delta^{1/2} F N^{-1} + \delta^{-1/2} Y + F^{-1/2} NL + YL.$$

The sum of the second and fourth terms is $\ll \delta^{-1/2} YL$, so the lemma follows.

In the final result of this section, we show that, in the language of Section 6.1, $(1/2, 1/2; 1/2, 1/2)$ is an exponent quadruple.

LEMMA 6.11. *Assume* $(\Omega_1), (\Omega_2)$, and (Ω_3). *Assume also that*

$$f_x \approx FX^{-1}, f_y \approx FY^{-1}, f_{xx} \approx FX^{-2}, f_{xy} \approx FY^{-2}, \text{ and}$$

$$Hf \approx F^2 X^{-2} Y^{-2}.$$

Let $XY = N$. *Then*

$$S \ll F + F^{-1/2} NL.$$

Proof. By the previous lemma,

$$S \ll F + F^{-1/2} NL + XL.$$

Interchanging the roles of X and Y yields

$$S \ll F + F^{-1/2} NL + YL.$$

By convexity,

$$S \ll F + F^{-1/2} NL + N^{1/2} L. \tag{6.3.6}$$

Now

$$\max(F, F^{-1/2} NL) \geq N^{2/3} L^{2/3} \gg N^{1/2} L,$$

so the last term in (6.3.6) may be omitted.

6.4 THE AB THEOREM

THEOREM 6.12. *Suppose $f(x,y) = Ax^{-\alpha}y^{-\beta}$, where α and β are real numbers with*

$$(\alpha)_3(\beta)_3(\alpha+\beta+1)_2 \neq 0.$$

Suppose $\mathbf{D} \subseteq (X,2X] \times (Y,2Y]$ is a domain such that conditions (Ω_2) and (Ω_3) are satisfied for f on $\mathbf{D}$. Let $F = AX^{-\alpha}Y^{-\beta}$, $N = XY$, and $L = \log(FN+2)$. Suppose further that $Y \leq X$. Then

$$S = \sum_{(m,n)\in\mathbf{D}} e(f(m,n)) \ll F^{1/3}N^{1/2} + N^{5/6}L^{2/3} + F^{-1/8}N^{15/16}L^{3/8} + F^{-1/4}NL^{1/2}.$$

Proof. Let Z be a parameter to be chosen later. For the moment, we assume only that $Z \leq N$. Let $Q = \sqrt{ZX/Y}$ and $R = \sqrt{ZY/X}$. Note that $Q/X = R/Y = \sqrt{Z/N}$. By the Weyl-van der Corput inequality ,

$$|S|^2 \ll \frac{N^2}{Z} + \frac{N}{Z} \sum_{(q,r)\in\mathbf{Q}} |S_1(q,r)|, \tag{6.4.1}$$

where

$$\mathbf{Q} = \{(q,r) : |q| \leq Q, |r| \leq R, (q,r) \neq (0,0)\},$$

$$S_1(q,r) = \sum_{(m,n)\in\mathbf{D}(q,r)} e(f_1(m,n;q,r)),$$

$$f_1(m,n;q,r) = \int_0^1 \frac{\partial}{\partial t} f(m+qt, n+rt)\, dt,$$

and

$$\mathbf{D}(q,r) = \{(m,n) : (m+qt, n+rt) \in \mathbf{D} \text{ for } t = 0,1\}.$$

For brevity, we write $f_1(m,n)$ in place of $f_1(m,n;q,r)$. Let $\rho = \max(|q|/X, |r|/Y)$. Assume for simplicity that $\rho = |r|/Y$; i.e. $|r|/Y \geq |q|/X$. (The contrary case can be similarly handled.) It follows that $r \neq 0$.

A straightforward calculation shows that

$$D_{x^a y^b} f_1(m,n) = (-1)^{a+b} A m^{-\alpha-a} n^{-\beta-b} \frac{r}{n} \{T_{a,b}(\frac{qn}{rm}) + O(\rho)\} \tag{6.4.2}$$

where

$$T_{a,b}(x) = (\alpha)_{a+1}(\beta)_b x + (\alpha)_a(\beta)_{b+1}.$$

Moreover,

$$Hf_1(m,n) = A^2 m^{-2\alpha-2} n^{-2\beta-2} \frac{r^2}{n^2} \{U(\frac{qn}{rm}) + O(\rho)\}, \tag{6.4.3}$$

where

$$U(x) = \begin{vmatrix} T_{2,0}(x) & T_{1,1}(x) \\ T_{1,1}(x) & T_{0,2}(x) \end{vmatrix} = \alpha\beta(\alpha+\beta+2)\{(\alpha)_2 x^2 + 2(\alpha+1)(\beta+1)x + (\beta)_2\}.$$

Note that $U(x)$ is not identically 0 since $\alpha\beta(\alpha+\beta+2) \neq 0$. Moreover, the roots of U are distinct since

$$-(\alpha)_2(\beta)_2 + (\alpha+1)^2(\beta+1)^2 = (\alpha+1)(\beta+1)(\alpha+\beta+1) \neq 0.$$

Finally, note that neither coefficient of $T_{0,2}$ or $T_{2,0}$ is 0. From equations (6.4.2), and (6.4.3), we see that there is a constant C such that

$$\left|D_{xx}f_1(m,n) + Am^{-\alpha-2}n^{-\beta}rn^{-1}T_{2,0}(qn/rm)\right| \leq C\rho^2 FX^{-2},$$

$$\left|D_{yy}f_1(m,n) + Am^{-\alpha}n^{-\beta-2}rn^{-1}T_{0,2}(qn/rm)\right| \leq C\rho^2 FY^{-2},$$

and

$$\left|Hf_1(m,n) - A^2m^{-2\alpha-2}n^{-2\beta-2}r^2n^{-2}U(qn/rm)\right| \leq C\rho^3 F^2X^{-2}Y^{-2}.$$

for all $(m,n) \in \mathbf{D}$.

Let δ be a small parameter to be chosen later, and let c be a small positive constant. Define

$$\mathbf{D}_0 = \{(m,n) \in \mathbf{D} : |Hf_1| \geq \delta\rho^2F^2N^{-2} \text{ and } |D_{uu}f_1| \geq c\rho FX^{-2}\},$$

$$\mathbf{D}_1 = \{(m,n) \in \mathbf{D} : |Hf_1| \geq \delta\rho^2F^2N^{-2} \text{ and } |D_{uu}f_1| < c\rho FX^{-2}\},$$

$$\mathbf{D}_2 = \{(m,n) \in \mathbf{D} : |Hf_1| < \delta\rho^2F^2N^{-2}\},$$

and define

$$T_\nu = \sum_{(m,n)\in\mathbf{D}_\nu} e(f_1(m,n;q,r))$$

for $\nu = 0, 1$, and 2.

By Lemma 6.10,

$$T_0 \ll \rho F + \rho^{-1/2}F^{-1/2}NL + \delta^{-1/2}YL.$$

By Lemma 6.5, $D_{yy}f_1 \gg \rho FY^{-2}$ and $Hf_1 \gg \rho FX^{-2}$ on $\mathbf{D}_1$. Consequently,

$$T_1 \ll \rho F + \rho^{-1/2}F^{-1/2}NL + XL$$

by Lemma 6.10.

By Lemma 6.4, the number of points in $\mathbf{D}_2$ is $\ll (\delta+\rho)N + 1$. From Lemma 6,5, we see that $D_{uu}f_1 \gg \rho FX^{-2}$ on $\mathbf{D}_2$. From Lemma 6.6,

$$T_2 \ll |\mathbf{D}_2|\rho^{1/2}F^{1/2}X^{-1} + \rho^{-1/2}F^{-1/2}N.$$

We use the above when $(\delta+\rho)N \geq 1$ and the trivial estimate otherwise to obtain

$$T_2 \ll (\delta+\rho)\rho^{1/2}F^{1/2}Y + \rho^{-1/2}F^{-1/2}N + 1.$$

Combining the estimates for T_0, T_1, and T_2, we find that

$$S_1(q,r) \ll \rho F + \rho^{-1/2}F^{-1/2}NL + \delta^{-1/2}YL + NY^{-1}L + (\delta+\rho)\rho^{1/2}F^{1/2}Y.$$

We take $\delta = c(\rho F)^{-1/3}$ for some c sufficiently small. Then

$$S_1(q,r) \ll \rho F + \rho^{-1/2}F^{-1/2}NL + \rho^{-1/6}F^{1/6}YL + NY^{-1}L + \\ + \rho^{3/2}F^{1/2}Y. \tag{6.4.4}$$

Now if $a > -1$ then

$$\frac{1}{Z}\sum_{(q,r)\in \mathbf{Q}} \rho^a \ll \frac{1}{Z}\sum_{1\leq q\leq Q}\sum_{1\leq r\leq R}\left(\frac{q^a}{X^a}+\frac{r^a}{Y^a}\right) \ll \left(\frac{Z}{N}\right)^{a/2}.$$

We combine this with (6.4.1) and (6.4.4) to get

$$S^2 \ll N^2Z^{-1} + FN^{1/2}Z^{1/2} + F^{-1/2}N^{9/4}Z^{-1/4}L + F^{1/6}N^{11/12}YZ^{1/12} + \\ + N^2Y^{-1}L + F^{1/2}N^{1/4}YZ^{3/4}.$$

Now we apply Lemma 2.4 with the condition $Z \leq N$; we then obtain

$$S^2 \ll F^{2/3}N + N^{5/3}L^{2/3} + F^{2/13}NY^{12/13}L^{12/13} + N^{5/4}Y^{3/4}L + \\ + F^{2/7}NY^{4/7} + F^{-1/4}N^{7/4}Y^{1/4}L^{3/4} + N^2Y^{-1}L + F^{-1/2}N^2L. \tag{6.4.5}$$

Since we are assuming that $Y \leq X$ and $XY = N$, we may assume that $Y \leq N^{1/2}$. Employing this in (6.4.5), we find that

$$S^2 \ll F^{2/3}N + N^{5/3}L^{2/3} + F^{2/13}N^{19/13}L^{12/13} + \\ F^{2/7}N^{9/7} + F^{-1/4}N^{15/8}L^{3/4} + N^2Y^{-1}L + F^{-1/2}N^2L. \tag{6.4.6}$$

Using convexity, we find that

$$\max(F^{2/3}N, F^{-1/2}N^2L) \geq F^{2/7}N^{9/7}$$

and

$$\max(F^{2/3}N, F^{-1/4}N^{15/8}L^{3/4}) \geq F^{2/13}N^{19/13}L^{12/13}.$$

Therefore

$$S^2 \ll F^{2/3}N + N^{5/3}L^{2/3} + F^{-1/4}N^{15/8}L^{3/4} + F^{-1/2}N^2L + N^2Y^{-1}L.$$

From Lemma 6.10, we see that $S^2 \ll FY^2$. Thus if $N^2Y^{-1}L$ is the maximum term in (6.4.6), then

$$S^2 \ll \min(N^2Y^{-1}L, FY^2) \leq F^{2/3}N^{4/3}L^{2/3} \leq \max(F^{2/3}N, N^{5/3}L^{4/3}).$$

Thus the $N^2Y^{-1}L$ term in (6.4.6) may the eliminated at the price of replacing $N^{5/3}L^{2/3}$ with $N^{5/3}L^{4/3}$.

7. NEW EXPONENT PAIRS

7.1 INTRODUCTION

In this section, we present a method of Bombieri and Iwaniec that leads to exponent pairs not obtainable from the A and B processes.

THEOREM 7.1. *For any* $\epsilon > 0$,

$$(\frac{9}{56}+\epsilon, \frac{37}{56}+\epsilon)$$

is an exponent pair.

Bombieri and Iwaniec (1986a,b) proved this theorem in the case $f(t) = t \log n$, and thereby proved that

$$\zeta(\frac{1}{2}+it) \ll t^{9/56+\epsilon}.$$

Huxley and Watt (1988) later extended the result to more general functions. The proof combines elements of van der Corput's method, Weyl's method, Vinogradov's mean value theorem, and the large sieve. As in Weyl's method, the function f is approximated by a polynomial; here the polynomial has degree 3. To evaluate the resulting sums requires an analysis of cubic exponential sums; this is given in Sections 7.3 and 7.4. Further analysis is accomplished by means of the elaborate version of the large sieve given in Lemma 7.5. This in turn requires upper bounds on certain sets of rational points (Lemma 7.19) and semicubical powers (Lemma 7.21).

In proving this theorem, we may assume that f is defined on some interval $[M, M_1]$, where $M_1 \le 2M$, and that for some y, s, P, and ϵ_0, the function f is an element of $\mathbf{F}(M, P, s, y, \epsilon_0)$. For convenience, we write

$$F = yM^{-s+1}.$$

We need to show that

$$\sum_{M<m\le M_1} e(f(m)) \ll F^{9/56+\epsilon}M^{1/2+\epsilon} + F^{-1}M.$$

As we remarked in Chapter 3, we may assume that $F \ge 1$.

We begin by eliminating the values of F that can be handled with classical exponent pairs. When $M < F^{3/7}$, we use the exponent pair $ABABAB(0,1) = (1/9, 13/18)$ to obtain

$$S \ll F^{1/9}M^{11/18} \ll F^{9/56}M^{1/2}.$$

When $M > F^{4/7}$, we use $BABABAB(0,1) = (2/9, 11/18)$; this yields

$$S \ll F^{2/9} M^{7/18} \ll F^{9/56} M^{1/2}.$$

For the remainder of the proof, we do not need the full strength of the exponent pair hypotheses. In fact, we need make only the following two assumptions.

1. f is defined and has four continuous derivatives on an interval $[M, M_1]$ with $M_1 \le 2M$.
2. There are constants c_1, c_2 such that
$$\frac{c_1 F}{M^j} \le (-1)^{j+1} f^{(j)}(x) \le \frac{c_2 F}{M^j} \tag{7.1.1}$$
for $j = 1, \ldots, 4$ and $x \in [M, M_1]$.

7.2 PRELIMINARIES

LEMMA 7.2. *Suppose that $N < M$ and that $\omega(x)$ is a real, continuously differentiable function defined on $[N, M]$. Suppose further that $|\omega(x)| \le \Omega_0$ and that $|\omega'(x)| \le \Omega_1$ on $[N, M]$. For any complex numbers a_n,*

$$\Big|\sum_{N<n\le M} a_n \omega(n)\Big| \le \Omega_0 \Big|\sum_{N<n\le M} a_n\Big| + \Omega_1 \int_N^M \Big|\sum_{N<n\le x} a_n\Big| dx$$

Proof. This is an easy consequence of integration by parts. Let

$$S(x) = \sum_{N<n\le x} a_n.$$

Then

$$\sum_{N<n\le M} a_n \omega(n) = \int_N^M \omega(x)\, dS(x) = S(x)\omega(x)\Big]_N^M - \int_N^M S(x)\omega'(x)\, dx,$$

and the stated result follows.

LEMMA 7.3. *Suppose $M \le N < N_1 \le M_1$. Let*

$$K(\theta) = \min\{M_1 - M + 1, (\pi|\theta|)^{-1}, (\pi\theta)^{-2}\}.$$

Then

$$\Big|\sum_{N<n\le N_1} a_n\Big| \le \int_{-\infty}^{\infty} K(\theta) \Big|\sum_{M<m\le M_1} a_m e(m\theta)\Big|\, d\theta.$$

Moreover, the L_1-norm of $K(\theta)$ is

$$\int_{-\infty}^{\infty} K(\theta)\, d\theta \ll \log(M_1 - M + 2).$$

Proof. The sum on the left-hand side is equal to

$$\sum_{M<m\le M_1} a_m \chi(m),$$

where $\chi(m)$ is the function with the graph

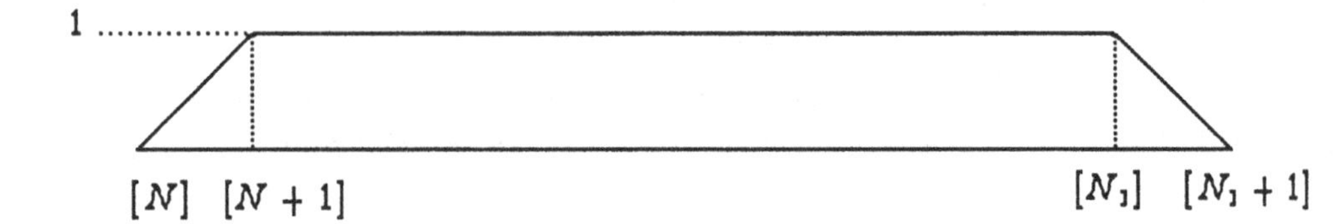

By the Fourier inversion formula,

$$\sum_{M<m\le M_1} a_m \chi(m) = \int_{-\infty}^{\infty} \hat{\chi}(\theta) \sum_{M<m\le M_1} a_m \chi(m) e(-m\theta)\, d\theta.$$

To complete the proof of the first statement, we need to show that $|\hat{\chi}(\theta)| \le K(\theta)$. We do this in the following steps. First, let R be a parameter to be chosen later and set $f_R(x) = \max(R - |x|, 0)$. Then

$$\hat{f}_R(\theta) = \frac{sin^2 \pi R\theta}{\pi^2\theta^2}.$$

Let $g(x) = f_R(x) - f_{R-1}(x)$. Then

$$\hat{g}(\theta) = \frac{\sin^2(\pi\theta R) - \sin^2(\pi\theta(R-1))}{\pi^2\theta^2} = -\frac{1}{\pi\theta}\int_{R-1}^{R} \sin 2\pi\theta x dx. \tag{7.2.1}$$

Now $\chi(x) = g(x - L)$, where $L = ([N_1 + 1] + [N])/2$ and $R = ([N_1 + 1] - [N])/2$. Therefore $\hat{\chi}(\theta) = \hat{g}(\theta)e(L\theta)$ and $|\hat{\chi}(\theta)| = |\hat{g}(\theta)|$. From (7.2.1), we see that

$$|\hat{\chi}(\theta)| \le \min\{(\pi|\theta|)^{-1}, (\pi\theta)^{-2}\}.$$

The inequality $|\hat{\chi}(\theta)| \le N_1 - N + 1$ is a trivial consequence of the definition of $\hat{\chi}(\theta)$. This completes the proof of $|\hat{\chi}(\theta)| \le K(\theta)$.

To evaluate the L_1-norm of K, let $\xi = 1/(M_1 - M + 1)$. Then

$$\int_{-\infty}^{\infty} K(\theta)\, d\theta \ll \int_0^{\xi} (M_1 - M + 1)\, d\theta + \int_{\xi}^{1} \theta^{-1}\, d\theta + \int_1^{\infty} \theta^{-2}\, d\theta \ll \log(M_1 - M + 2).$$

LEMMA 7.4. *Let* $\mathbf{P}$ *be a set of points* $\mathbf{p} \in \mathbf{R}^K$ *and let* $b(\mathbf{p})$*, for* $\mathbf{p} \in \mathbf{P}$*, be arbitrary complex numbers. Let* $\delta_1, \ldots, \delta_K, T_1, \ldots, T_K$ *be positive numbers. Then*

$$\int_{-T_1}^{T_1} \cdots \int_{-T_K}^{T_K} \Big|\sum_{\mathbf{p}\in\mathbf{P}} b(p) e(\mathbf{p}\cdot\mathbf{t})\Big|^2 dt_1 \ldots dt_K \le \prod_{k=1}^{K} (2T_k + \delta_k^{-1}) \sum_{\substack{\mathbf{p}\in\mathbf{P},\, \mathbf{p}'\in\mathbf{P} \\ |p_j - p_j'| \le \delta_j}} |b(\mathbf{p})b(\mathbf{p}')|.$$

Proof. Let δ and T be positive numbers. Using the Beurling-Selberg function of the Appendix, we can construct a function $f(t)$ such that $f(t) \geq 1$ if $|t| \leq T$, $f(t) \geq 0$ for all t, the Fourier transform $\hat{f}(u) = 0$ if $u \geq \delta$, and $\hat{f}(0) = 2T + \delta^{-1}$. Since

$$\int_{-T_1}^{T_1} \cdots \int_{-T_K}^{T_K} \Big|\sum_{\mathbf{p}\in\mathbf{P}} b(\mathbf{p})e(\mathbf{p}\cdot\mathbf{t})\Big|^2 dt_1 \ldots dt_K \leq$$
$$\int_{-\infty}^{\infty} \cdots \int_{-\infty}^{\infty} f_{\delta_1}(t_1)\ldots f_{\delta_K}(t_K)\Big|\sum_{\mathbf{p}\in\mathbf{P}} b(\mathbf{p})e(\mathbf{p}\cdot\mathbf{t})\Big|^2 dt_1 \ldots dt_K,$$

the result follows.

LEMMA 7.5. *Let* $\mathbf{X}$ *and* $\mathbf{Y}$ *be two subsets of* $\mathbf{R}^K$. *Let* $a(\mathbf{x})$ *and* $b(\mathbf{y})$ *be arbitrary complex numbers for* $\mathbf{x} \in \mathbf{X}$ *and* $\mathbf{y} \in \mathbf{Y}$. *Let* $X_1, \ldots, X_K, Y_1, \ldots, Y_K$ *be positive numbers. Define the bilinear forms*

$$\mathbf{B}(b;\mathbf{X}) = \sum_{\mathbf{y}\in\mathbf{Y}} \sum_{\substack{\mathbf{y}'\in\mathbf{Y} \\ |y_k - y'_k| \leq (2X_k)^{-1};\ k=1,\ldots,K}} |b(\mathbf{y})b(\mathbf{y}')|,$$

$$\mathbf{B}(a;\mathbf{X}) = \sum_{\mathbf{x}\in\mathbf{X}} \sum_{\substack{\mathbf{x}'\in\mathbf{X} \\ |x_k - x'_k| \leq (2Y_k)^{-1};\ k=1,\ldots,K}} |a(\mathbf{x})a(\mathbf{x}')|,$$

$$\mathbf{B}(a,b;\mathbf{X},\mathbf{Y}) = \sum_{\substack{\mathbf{x}\in\mathbf{X} \\ |x_k|\leq X_k}} \sum_{\substack{\mathbf{y}\in\mathbf{Y} \\ |y_k|\leq Y_k}} a(\mathbf{x})b(\mathbf{y})e(\mathbf{x}\cdot\mathbf{y}).$$

Then

$$|\mathbf{B}(a,b;\mathbf{X},\mathbf{Y})|^2 \leq (2\pi^2)^K \prod_{k=1}^{K}(1 + X_kY_k)\mathbf{B}(b;\mathbf{X})\mathbf{B}(a;\mathbf{Y}).$$

Proof. We begin with the integral formula

$$e(xy) = \frac{\pi y}{\sin(2\pi\epsilon y)} \int_{x-\epsilon}^{x+\epsilon} e(ty)\,dt.$$

We set $\epsilon_k = (4Y_k)^{-1}$ and

$$w(\mathbf{y}) = \prod_{k=1}^{K} \frac{\pi y_k}{\sin(2\pi\epsilon_k y_k)}.$$

Then for $\mathbf{y}$ with $|y_k| \leq Y_k$, we have

$$w(\mathbf{y}) \leq \pi^K Y_1 \ldots Y_K.$$

We see that if $b^*(\mathbf{y}) = b(\mathbf{y})w(\mathbf{y})$, then

$$|\mathbf{B}(a,b;\mathbf{X},\mathbf{Y})| = \Big|\sum_{\mathbf{x}\in\mathbf{X}} a(x) \int_{x-\epsilon_1}^{x+\epsilon_1} \cdots \int_{x-\epsilon_K}^{x+\epsilon_K} \sum_{\mathbf{y}\in\mathbf{Y}} b^*(\mathbf{y})e(\mathbf{t}\cdot\mathbf{y})\,dt_1 \ldots dt_K\Big|.$$

After setting $T_k = \epsilon_k + X_k$ and using Cauchy's inequality, we obtain

$$\leq \int_{-T_1}^{T_K} \cdots \int_{-T_K}^{T_K} \sum_{\substack{\mathbf{x}\in\mathbf{X} \\ |x_k - t_k|\leq \epsilon_k}} |a(x)| \, \Big|\sum_{\mathbf{y}\in\mathbf{Y}} b^*(\mathbf{y}) e(\mathbf{t}\cdot\mathbf{y})\Big| \, dt_1 \ldots dt_K$$

$$\leq \left(\int \cdots \int |\sum_{\mathbf{X}}|^2\right)^{1/2} \left(\int \cdots \int |\sum_{\mathbf{Y}}|^2\right)^{1/2}.$$

From Lemma 7.4 we see that

$$\int \cdots \int |\sum_{\mathbf{X}}|^2 \leq 2^K \epsilon_1 \ldots \epsilon_K \sum_{\substack{\mathbf{x},\mathbf{x}'\in\mathbf{X} \\ |x_k - x'_k|\leq 2\epsilon_k}} |a(x)a(x')|$$

and

$$\int \cdots \int |\sum_{\mathbf{Y}}|^2 \leq \prod_{k=1}^{K} (2T_k + \delta_k^{-1}) \sum_{\substack{\mathbf{y}\in\mathbf{Y},\mathbf{y}'\in\mathbf{Y} \\ |y_k - y'_k|\leq \delta_k}} |b^*(\mathbf{y}) b^*(\mathbf{y}')|$$

for any $\delta_k > 0$. Choosing $\delta_k = (2X_k)^{-1}$ completes the proof.

In our applications, we will also need the following simple modification of the previous result. Suppose that for some k's, the x_k's all assume integral values. Then the norms $|y_k|$ and $|y_k - y'_k|$ can be replaced by $||y_k||$ and $||y_k - y'_k||$ respectively if we set $Y_k = 1$ for such k's.

7.3 THE AIRY-HARDY INTEGRAL

LEMMA 7.6. *(Hardy, 1910) If $y > 0$ then*

$$\int_0^\infty e(x^3 - 3yx)\, dx = \frac{e(1/8 - 2y^{3/2})}{6^{1/2} y^{1/4}} + O\left(\frac{1}{y}\right).$$

Proof. For $0 < y < 1$ the assertion is a straightforward consequence of Lemma 3.1 and the trivial estimate.

Now suppose $y \geq 1$. We have

$$x^3 - 3yx = -2y^{3/2} + 3y^{1/2}(x - y^{1/2})^2 + (x - y^{1/2})^3.$$

Making the substitution $t = (x - y^{1/2})y^{-1}$, we find that

$$\int_0^\infty e(x^3 - 3yx)\, dx = e(-2y^{3/2}) y^{1/2} \int_{-1}^\infty e(y^{3/2}(3t^2 + t^3))\, dt$$

Since

$$\int_1^\infty e(y^{3/2}(3t^2+t^3))\,dt \ll y^{-3/2}$$

by Lemma 3.1, we may write

$$\int_0^\infty e(x^3-3yx)\,dx = e(-2y^{3/2})y^{1/2}(I_1+I_2)+O\left(y^{-1}\right), \tag{7.3.1}$$

where

$$I_1=\int_0^1 e(y^{3/2}(3t^2+t^3))\,dt \text{ and } I_2=\int_0^1 e(y^{3/2}(3t^2-t^3))dt.$$

By the change of variable $3t^2+t^3=u$, we get

$$I_1=\int_0^4 e(y^{3/2}u)(6t+3t^2)^{-1}\,du.$$

Now we note that

$$(6t+3t^2)^{-1}=\frac{1}{2(3u)^{1/2}}+f(u),$$

where $f(u)=c_0+c_1u^{1/2}+c_2u+\ldots$ and $f(z^2)$ is analytic in $|z|\le 4$. Consequently,

$$I_1=\frac{1}{2\cdot 3^{1/2}}\int_0^4 e(y^{3/2}u)u^{-1/2}\,du+O(y^{-3/2}).$$

Making the change of variable $u=v^2$ and using Lemma 3.3, we find that

$$I_1=\frac{1}{2\cdot 6^{1/2}}\frac{e(1/8)}{y^{3/4}}+O(y^{-3/2}).$$

In a similar fashion, we may show that the above is also true when I_1 is replaced by I_2. The result now follows from (7.3.1).

An immediate consequence of the preceding lemma is that if μ, c, and h are all positive then

$$\int_0^\infty e\left(\mu x^3-\frac{h}{cx}\right)dx=\left(\frac{c}{12\mu h}\right)^{1/4}e(1/8)e\left(-2\mu^{-1/2}\left(\frac{h}{3c}\right)^{3/2}\right)+O\left(\frac{c}{h}\right). \tag{7.3.2}$$

LEMMA 7.7. *Suppose $\mu>0, c\ge 1$, and $1\le N\le N_1\le 2N$. Let $\delta(h)$ be the characteristic function of the interval $[3\mu cN^2, 3\mu cN_1^2]$. If h is real and non-zero, then*

$$\int_N^{N_1} e\left(\mu x^3-\frac{h}{cx}\right)dx=\delta(h)\left(\frac{c}{12\mu h}\right)^{1/4}e(1/8)e\left(-2\mu^{-1/2}\left(\frac{h}{3c}\right)^{3/2}\right)$$
$$+O\left(\min\left\{(\mu N)^{-1/2},\left|3\mu N^2-\frac{h}{c}\right|^{-1}+\left|3\mu N_1^2-\frac{h}{c}\right|^{-1}\right\}\right)+O(c|h|^{-1}).$$

Proof. When $\delta(h)=1$, this follows by using (7.3.2) together with Lemmas 3.1 and 3.2 to estimate the tails. When $\delta(h)=0$, this follows from Lemmas 3.1 and 3.2.

LEMMA 7.8. *Let $g(x)$ be the function whose graph is given below.*

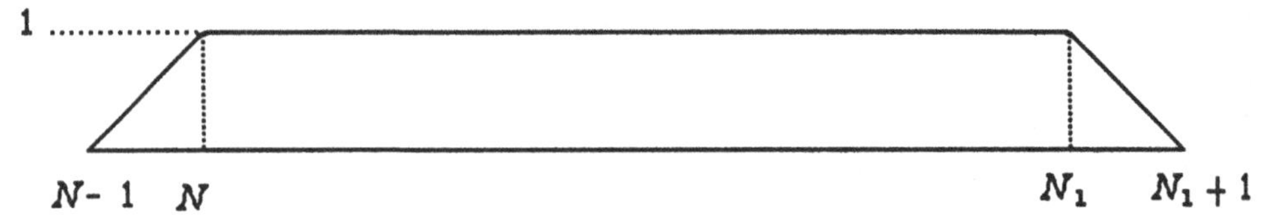

Suppose that either $h < 2\mu c N^2$ or $h > 2\mu c N_1^2$. Then

$$\int g(x) e\left(\mu x^3 - \frac{h}{cx}\right) dx \ll c(\mu c N^2 + |h|)^{-1}.$$

The implied constant is absolute.

Proof. Let

$$M(x) = \frac{cg(x)}{3\mu c x^2 - h}.$$

By partial integration, the integral is equal to

$$\frac{-1}{2\pi i}\int M'(x) e\left(\mu x^3 - \frac{h}{cx}\right) dx \ll \int |M'(x)|\, dx.$$

Now when x is not a point of discontinuity of $g'(x)$, we may write

$$M''(x) = \frac{6\mu c^2(9\mu c x^2 + h)g(x) - 12\mu c^2 x(3\mu c x^2 - h)g'(x)}{(3\mu c x^2 - h)^3}.$$

Since the numerator is piecewise polynomial, $M'(x)$ is monotonic on a finite number of subintervals. Therefore

$$\int |M'(x)| dx \ll \max |M(x)| \ll c(\mu c N^2 + |h|)^{-1}.$$

LEMMA 7.9. *If $|h| > 4\mu c N_1^2$ then*

$$\int g(x) e\left(\mu x^3 - \frac{h}{cx}\right) dx \ll (1 + \mu N^2)c^2 h^{-2}.$$

The implied constant is absolute.

Proof. Use the formula from the proof of the previous integral, and write the integral as $I_1 + I_2 - I_3$, where

$$I_1 = \frac{6\mu c^2}{2\pi i}\int \frac{xg(x)}{(3\mu c x^2 - h)^2} e\left(\mu x^3 - \frac{h}{cx}\right) dx,$$

$$I_2 = \frac{1}{2\pi i}\int_N^{N+1} \frac{c}{3\mu c x^2 - h} e\left(\mu x^3 - \frac{h}{cx}\right) dx,$$

$$I_3 = \frac{1}{2\pi i}\int_{N_1}^{N_1+1} \frac{c}{3\mu c x^2 - h} e\left(\mu x^3 - \frac{h}{cx}\right) dx,$$

From the trivial estimate, we see that $I_1 \ll \mu c^2 N^2 h^{-2}$. From Theorem 3.1, we see that $I_2 \ll c^2 h^{-2}$, and the same estimate holds for I_3.

7.4 GAUSS SUMS

Define the Gauss sums

$$G(a,l;c) = \sum_{d \bmod c} e\left(\frac{ad^2+ld}{c}\right). \tag{7.4.1}$$

In this section, we shall show that if $(a,c)=1$ then

$$|G(a,l;c)| \le (2c)^{1/2}. \tag{7.4.2}$$

We will then use the above to obtain an estimate for cubic exponential sums.

To avoid repetition, we shall implicitly assume that $(a,c)=1$ in any discussed sum of the form $G(a,l;c)$.

LEMMA 7.10. *If $(m,n)=1$ then*

$$G(a,l;mn) = G(an,l;m)G(am,l;n).$$

Proof. The left-hand side is

$$\sum_{d \bmod mn} e\left(\frac{ad^2+ld}{mn}\right) = \sum_{j \bmod m}\sum_{k \bmod n} e\left(\frac{a(jn+km)^2+l(jn+km)}{mn}\right)$$
$$= \sum_{j \bmod m} e\left(\frac{anj^2+lj}{m}\right)\sum_{k \bmod n} e\left(\frac{amk^2+lk}{n}\right),$$

and the result follows.

In the next lemma, we use $\bar{a}$ to denote the multiplicative inverse of a mod c. In other words, $\bar{a}$ is defined by $a\bar{a} \equiv 1 \bmod c$.

LEMMA 7.11.

(a) *If c is odd then*

$$G(a,l;c) = e\left(-\frac{4\bar{a}}{c}\frac{l^2}{4}\right)G(a,0;c).$$

(b) *If l is even then*

$$G(a,l;c) = e\left(-\frac{\bar{a}}{c}\frac{l^2}{4}\right)G(a,0;c).$$

(c) *If l is odd then*

$$G(a,l;c) = e\left(-\frac{\bar{a}}{c}\frac{l^2-1}{4}\right)G(a,1;c).$$

Proof. The first statement follows readily after replacing d by $d+2\bar{a}l$ in the index of summation of (7.4.1). Similarly, the second statement follows upon replacing d by $d+\bar{a}l/2$, and the third statement follows upon replacing d by $d+\bar{a}(l-1)/2$.

LEMMA 7.12. *If p is an odd prime and $\alpha \geq 2$ then $G(a,0;p^\alpha) = pG(a,0;p^{\alpha-2})$.*

Proof.

$$G(a,0;p^\alpha) = \sum_{j=0}^{p-1}\sum_{k=1}^{p^{\alpha-1}} e\left(\frac{a(jp^{\alpha-1}+k)^2}{p^\alpha}\right) = \sum_{k=1}^{p^{\alpha-1}} e\left(\frac{ak^2}{p^\alpha}\right)\sum_{j=0}^{p-1} e\left(\frac{2ajk}{p}\right).$$

Since the inner sum is p if $p|k$ and 0 otherwise, the result follows.

An immediate consequence of the previous result is that

$$G(a,0;p^\alpha) = \begin{cases} p^{\alpha/2} & \text{if } \alpha \text{ even;} \\ p^{(\alpha-1)/2}G(a,0;p) & \text{if } \alpha \text{ odd.} \end{cases}$$

In the next lemma, we use $\left(\frac{a}{p}\right)$ to denote the Legendre symbol of a mod p.

LEMMA 7.13. *If p is an odd prime, then*

$$G(a,0;p) = \left(\frac{a}{p}\right)G(1,0;p).$$

Proof. Suppose $1 \leq k \leq p$. The number of solutions of $ad^2 \equiv k \bmod p$ is $1 + \left(\frac{ak}{p}\right)$, so

$$\sum_{d \bmod p} e\left(\frac{ad^2}{p}\right) = \sum_{k \bmod p} e\left(\frac{k}{p}\right)\left(1 + \left(\frac{ak}{p}\right)\right) = \left(\frac{a}{p}\right)\sum_{k \bmod p} e\left(\frac{k}{p}\right)\left(\frac{k}{p}\right)$$
$$= \left(\frac{a}{p}\right)G(1,0;p).$$

LEMMA 7.14. *Suppose that $c = 2^r$ and that l is odd. If $r = 1$ then $G(a,l;c) = 2$. If $r \geq 2$ then $G(a,l;c) = 0$.*

Proof. The first assertion is trivial. For the second assertion, we observe that

$$G(a,l;2^r) = \sum_{j=0}^{1}\sum_{k=1}^{2^{r-1}} e\left(\frac{a(j2^{r-1}+k)^2 + l(j2^{r-1}+k)}{2^r}\right)$$
$$= \sum_{k=1}^{2^{r-1}} e\left(\frac{ak^2+lk}{2^r}\right)\sum_{j=0}^{1} e\left(\frac{lj}{2^r}\right).$$

Since l is odd, the inner sum is 0.

LEMMA 7.15. *For any positive c,*

$$G(1,0;c) = \frac{1}{2}(1+i^{-c})(1+i)c^{1/2}.$$

Proof. By the Poisson summation formula,

$$\sum_{d=0}^{c-1} e\left(\frac{d^2}{c}\right) = \sum_{\nu=-\infty}^{\infty} \int_0^c e\left(\nu x + \frac{x^2}{c}\right) dx.$$

With a couple of changes of variable, we see that this is

$$= c\sum_{\nu=-\infty}^{\infty} \int_0^1 e(c(x+\nu x))\, dx = c\sum_{\nu=-\infty}^{\infty} e\left(-\frac{c\nu^2}{4}\right) \int_{\nu/2}^{1+\nu/2} e(cy^2)\, dy.$$

After breaking this last sum into odd and even terms, we see that it is

$$c(1+i^{-c}) \int_{-\infty}^{\infty} e(cy^2)\, dy.$$

By Lemma 3.3, the last integral is $e(1/8)(2c)^{-1/2}$, and the desired result follows.

We can now prove (7.4.2). If p is an odd prime, then $|G(a,l;p^\alpha)| = p^{\alpha/2}$ by Lemmas 7.11(a), 7.12, 7.13, and 7.15. Furthermore, $|G(a,l;2^\alpha)| \le 2^{(\alpha+1)/2}$ by Lemmas 7.11(b), 7.14, and 7.15. This together with Lemma 7.10 proves (7.4.2).

Now we are ready to evaluate the incomplete perturbed Gauss sum

$$S(\mu;a,b;c) = \sum_{N<n\le N_1} e\left(\mu n^3 + \frac{a}{cn^2} + \frac{b}{c}n\right).$$

LEMMA 7.16. *Suppose $c \ge 1, (a,c)=1, c \le N \le N_1 \le 2N$ and $0 < \mu \le N^{-2}$. Then $|S(\mu;a,b;c)| \le$*

$$\sum_{w\in\{-1,1\}} \left| \sum_{3\mu cN^2<h<3\mu cN_1^2} \frac{w^h}{(3\mu ch)^{1/4}} e\left(2\mu^{-1/2}\left(\frac{h}{3c}\right)^{3/2} + \frac{\bar{a}}{c}\left(\frac{b+h}{2}\right)^2\right)\right|$$

$$+ O(N^{1/2}\log N + \mu^{-1}N^{-2}).$$

The implied constant is absolute.

Proof. We begin by writing

$$\begin{aligned}
S(\mu;a,b;c) &= \sum_n g(n) e\left(\mu n^3 + \frac{a}{c}n^2 + \frac{b}{c}n\right) + O(1) \\
&= \sum_{d \bmod c} e\left(\frac{ad^2+bd}{c}\right) \sum_{n\equiv d \bmod c} g(n)e(\mu n^3) + O(1) \\
&= \frac{1}{c}\sum_{d \bmod c} e\left(\frac{ad^2+bd}{c}\right) \sum_n g(n)e(\mu n^3) \sum_{k \bmod c} e\left(\frac{(d-n)k}{c}\right) + O(1) \\
&= \frac{1}{c}\sum_h G(a,b+h;c) \int g(x)e\left(\mu x^3 - \frac{h}{c}x\right) dx + O(1),
\end{aligned}$$

where we have used the Poisson summation formula in the last line. The terms with $|h| \geq 16c$ contribute $\ll c^{1/2} \ll N^{1/2}$ by Lemma 7.9. By Lemma 7.8, the terms with $1 \leq |h| < 16c$ and $h \notin [2\mu c N^2, 4\mu c N_1^2]$ contribute $\ll c^{1/2} \log N \ll N^{1/2} \log N$. Similarly, the term with $h = 0$ contributes $\ll \mu^{-1} c^{-1/2} N^{-2} \ll \mu^{-1} N^{-2}$. It remains to consider the terms with $2\mu c N^2 < h < 4\mu c N_1^2$. If $16\mu c N^2 < 1$ then there are no such terms, and the lemma is proved. When $16\mu c N^2 \geq 1$, we write

$$\int g(x) e\left(\mu x^3 - \frac{h}{c} x\right) dx = \int_N^{N_1} e\left(\mu x^3 - \frac{h}{c} x\right) dx + O(1),$$

and we appeal to Lemma 7.7. The term attached to $\delta(h)$ gives rise to the main terms, and the total error term is

$$\begin{aligned} &\ll c^{-1/2} \sum_{2\mu c N^2 < h < 4\mu c N_1^2} (1 + ch^{-1}) \\ &\quad + c^{-1/2} \sum_{2\mu c N^2 < h < 4\mu c N_1^2} \min\left\{ (\mu N)^{-1/2}, \left|3\mu c N^2 - \frac{h}{c}\right|^{-1} + \left|3\mu c N_1^2 - \frac{h}{c}\right|^{-1} \right\} \\ &\ll \mu c^{1/2} N^2 + c^{1/2} \log N + (\mu c N)^{-1/2} \ll N^{1/2} \log N. \end{aligned}$$

7.5 LEMMAS ON RATIONAL POINTS

LEMMA 7.17. *Suppose r/q and r_1/q_1 are rationals in lowest terms such that*

$$\left|\frac{\bar{r}}{q} - \frac{\bar{r}_1}{q_1}\right| \leq \Delta,$$

where $\bar{r}, \bar{r}_1$ are defined by $r\bar{r} \equiv 1 \bmod q$ and $r_1 \bar{r}_1 \equiv 1 \bmod q_1$. Then there are integers $a, b, c,$ and d such that $ad - bc = 1$,

$$\begin{bmatrix} r_1 \\ q_1 \end{bmatrix} = \begin{bmatrix} a & b \\ c & d \end{bmatrix} \begin{bmatrix} r \\ q \end{bmatrix},$$

$$a = \frac{cr_1}{q_1} + \frac{q}{q_1}, d = -\frac{cr}{q} + \frac{q_1}{q}, b = \frac{-crr_1 + r_1 q_1 - rq}{qq_1},$$

and

$$|c| \leq \Delta q q_1.$$

Proof. Let $\bar{q}, \bar{q}_1$ be defined by $q\bar{q} + r\bar{r} = 1$, $q_1 \bar{q}_1 + r_1 \bar{r}_1 = 1$. Then

$$\begin{bmatrix} r_1 \\ q_1 \end{bmatrix} = \begin{bmatrix} r_1 & -\bar{q}_1 \\ q_1 & \bar{r}_1 \end{bmatrix} \begin{bmatrix} \bar{r} & \bar{q} \\ -q & r \end{bmatrix} \begin{bmatrix} r \\ q \end{bmatrix} = \begin{bmatrix} r_1 \bar{r} + q\bar{q}_1 & r_1 \bar{q} - r\bar{q}_1 \\ q_1 \bar{r} - q\bar{r}_1 & q_1 \bar{q} + r\bar{r}_1 \end{bmatrix} \begin{bmatrix} r \\ q \end{bmatrix}.$$

Thus we may take

$$\begin{bmatrix} a & b \\ c & d \end{bmatrix} = \begin{bmatrix} r_1 & -\bar{q}_1 \\ q_1 & \bar{r}_1 \end{bmatrix} \begin{bmatrix} \bar{r} & \bar{q} \\ -q & r \end{bmatrix}. \tag{7.5.1}$$

Both matrices on the right hand side have determinant 1, so $ad - bc = 1$. Since

$$c = (\bar{r}/q - \bar{r}_1/q_1)qq_1,$$

we see that $|c| \leq \Delta qq_1$. A straightforward calculation shows that

$$bqq_1 + crr_1 = r_1q_1 - rq;$$

this yields the stated expression for b. To get the stated expression for a, we premultiply both sides of (7.5.1) by $[q_1 \quad -r_1]$ to get $q = aq_1 - cr_1$. Similarly, postmultiplying both sides by $[r \quad q]^T$ gives $cr + dq = q_1$.

LEMMA 7.18. *Suppose A and C are positive integers, and that Δ_1 and Δ_2 are positive real numbers not exceeding 1. Suppose $h(x)$ is a real positive continuously differentiable function defined on a subinterval I of $[A/2C, 2A/C]$. Suppose that there is a constant C_0 and and a parameter H such that*

$$\frac{H}{C_0} \leq h(x) \leq C_0H\,, \frac{H}{C_0} \leq |xh'(x)| \leq C_0H\,, \frac{H}{C_0} \leq |h(x) - xh'(x)| \leq C_0H$$

whenever x is an element of I. Let B be the number of solutions of the inequalities

$$||\frac{\bar{r}}{q} - \frac{\bar{r}_1}{q_1}|| \leq \Delta_1, \tag{7.5.2}$$

$$|qh(\frac{r}{q}) - q_1h(\frac{r_1}{q_1})| \leq CH\Delta_2, \tag{7.5.3}$$

$$(r,q) = 1, (r_1,q_1) = 1, A < r, r_1 \leq 2A, C < q, q_1 \leq 2C. \tag{7.5.4}$$

Then

$$B \ll \Delta_1\Delta_2A^2C^2 + \Delta_1^2A^2C^2 + AC + \Delta_2A^2 + \Delta_2C^2.$$

The implied constant depends only on C_0.

Proof. For each pair of rationals $r/q, r_1/q_1$ satisfying (7.5.2) through (7.5.4), we can use Lemma 7.17 to find integers a, b, c, d such that

$$\begin{bmatrix} r_1 \\ q_1 \end{bmatrix} = \begin{bmatrix} a & b \\ c & d \end{bmatrix} \begin{bmatrix} r \\ q \end{bmatrix}. \tag{7.5.5}$$

Our proof proceeds by counting the number of matrices for which statements (7.5.2) through (7.5.5) are true. For convenience, we introduce the following notation. Whenever P is a statement, $B(P)$ denotes the number of pairs of rationals satisfying (7.5.2) through (7.5.5) and P. For example, $B(c = 0)$ is the number of solutions of (7.5.2) through (7.5.4) in which the matrix of (7.5.5) has $c = 0$.

We begin by observing that

$$B \leq B(a = 0) + B(b = 0) + B(c = 0) + B(d = 0) + B(abcd \neq 0).$$

If $c = 0$, then $q = q_1$ and $a = d = 1$. Moreover,

$$|r_1 - r|HA^{-1} \ll \Big|\int_{r/q}^{r_1/q} h'(u)\,du\Big| \ll H\Delta_2,$$

so

$$|bq| = |r_1 - r| \ll \Delta_2 A.$$

Thus for each rational r/q there are $\ll 1 + \Delta_2 AC^{-1}$ values of b, and so

$$B(c = 0) \ll AC(1 + \Delta_2 AC^{-1}) \ll AC + \Delta_2 A^2.$$

If $b = 0$ then $c = q_1/r - q/r_1$, $a = r_1/r$, and $d = r/r_1$. Moreover,

$$|q_1 - q|HA^{-1} \ll \Big|\int_{r/q_1}^{r/q} \frac{d}{du}\frac{h(u)}{u}\,du\Big| \ll \Delta_2 A^{-1}CH,$$

so

$$|cr| = |q_1 - q| \ll \Delta_2 C.$$

Thus for each rational r/q there are $\ll 1 + \Delta_2 A^{-1}C$ values of c, and

$$B(b = 0) \ll AC(1 + \Delta_2 A^{-1}C) \ll AC + \Delta_2 C^2.$$

If $a = 0$, then $c = -q/r_1 = -1$ and $b = 1$. Therefore $q = r_1$ and

$$d = \frac{r}{r_1} + \frac{q_1}{q}.$$

From condition (7.5.4), we see that $d \leq 4$. For each rational r/q there are at most four choices of d, and so

$$B(a = 0) \ll AC.$$

If $d = 0$, then $c = q_1/r = 1$ and $b = -1$. Therefore $q_1 = r$ and

$$a = \frac{r_1}{r} + \frac{q}{q_1}.$$

From condition (7.5.4), we see that $a \leq 4$, and

$$B(d = 0) \ll AC.$$

To complete the proof, we need to bound $B(abcd \neq 0)$, which we denote by B_1. We define $B_1(P)$ in a manner analogous to $B(P)$; i.e. $B_1(P) = B(abcd \neq 0 \text{ and } P)$.

Let D be an absolute constant to be chosen later. Then

$$B_1 = B_1(|bc| < D) + B_1(|bc| \geq D)$$

Now the number of matrices with $ad - bc = 1$ and $|bc| < D$ is

$$\ll 1 + \sum_{1<n<D+1} \{d(n)d(n+1) + d(n)d(n-1)\}.$$

From the trivial estimate $d(n) \leq n$, we see that the above is

$$\ll \sum_{n<D+1} n^2 \ll D^3;$$

whence

$$B_1(|bc| < D) \ll ACD^3 \ll AC.$$

Now suppose that $|bc| \geq D$. Then

$$D \leq |bc| = |\frac{-c^2 rr_1 + cr_1 q_1 - crq}{qq_1}| \leq \frac{4|c|^2 A^2}{C^2} + \frac{3|c|A}{C}.$$

If we let $w = |c|A/C$ and if we assume that $D \geq 7$, then $4w^2 + 3w \geq 7$, so $w \geq 1$. Therefore $w^2 \geq w$, $7w^2 \geq D$ and

$$|c| \geq \sqrt{D/7}CA^{-1}. \tag{7.5.6}$$

Furthermore,

$$\frac{A}{C}\left(\frac{|c|A}{4C} - 3\right) \leq |b| = |\frac{-crr_1 + r_1 q_1 - rq}{qq_1}| \leq \frac{A}{C}\left(\frac{4|c|A}{C} + 3\right).$$

If D is sufficiently large ($D \geq 10^6$ will do) then

$$\frac{|c|A^2}{5C^2} \leq |b| \leq \frac{5|c|A^2}{C^2}. \tag{7.5.7}$$

We also see that

$$\frac{|c|A}{5C} \leq \frac{|c|A}{4C} - 2 \leq |a| = |\frac{cr_1 + q}{q_1}| \leq \frac{4|c|A}{C} + 2 \leq \frac{5|c|A}{C} \tag{7.5.8}$$

and

$$\frac{|c|A}{5C} \leq |d| \leq \frac{5|c|A}{C}. \tag{7.5.9}$$

Now let $x = r/q$. Then

$$\frac{r_1}{q_1} = \frac{ax+b}{cx+d} \text{ and } \frac{q_1}{q} = cx + d.$$

If we let

$$g(x) = (cx+d)h\left(\frac{ax+b}{cx+d}\right) - h(x)$$

then we see that the inequality

$$|g(x)| \leq 2H\Delta_2$$

is a consequence of (7.5.3). Now

$$g'(x) = ch\left(\frac{ax+b}{cx+d}\right) + \frac{1}{cx+d}h'\left(\frac{ax+b}{cx+d}\right) - h'(x).$$

From our hypotheses on h and h', we see that

$$\left|\frac{1}{cx+d}h'\left(\frac{ax+b}{cx+d}\right) - h'(x)\right| \leq \frac{6C_0HC}{A}$$

and

$$\left|ch\left(\frac{ax+b}{cx+d}\right)\right| \geq \frac{|c|H}{C_0}.$$

From (7.5.6), we see that if D is sufficiently large ($D \geq 2000C_0^2$ will do) then

$$|g'(x)| \geq \frac{|c|H}{2C_0}. \tag{7.5.10}$$

We now choose D so that all of (7.5.6) through (7.5.10) are true; i.e. we take

$$D = \max(10^6, 2000C_0^2).$$

Given a matrix

$$\begin{bmatrix} a & b \\ c & d \end{bmatrix}$$

and a solution $r/q, r_1/q_1$ of (7.5.2) through (7.5.4), we have

$$|g(r/q)| \leq 2H\Delta_2.$$

Given the same matrix and a second solution $r'/q', r_1'/q_1'$, we also have

$$|g(r'/q')| \leq 2H\Delta_2.$$

Now

$$\frac{|c|H}{2C_0}\left|\frac{r}{q} - \frac{r'}{q'}\right| \leq \left|\int_{r'/q'}^{r/q} g'(u)\,du\right| \leq 4H\Delta_2,$$

so

$$|\frac{r}{q} - \frac{r'}{q'}| \leq \frac{8C_0\Delta_2}{|c|}.$$

We may therefore write

$$\frac{r}{q} - \frac{r'}{q'} = \frac{k}{qq'} \tag{7.5.11}$$

for some k with $|k| \leq 32C_0\Delta_2|c|^{-1}C^2$.

We claim that if r, q, and k are fixed, then there is at most one rational r'/q' for which (7.5.3) and (7.5.11) are true. For if r'/q' and r''/q'' are solutions, then

$$r(q' - q'') = q(r' - r'').$$

Since $(r, q) = 1$, we must have $q|(q' - q'')$. But $q \geq C$ and $|q' - q''| < C$, so $q' = q''$. Similarly, $r' = r''$. It follows that for fixed a, b, c, and d with $|bc| \geq D$ and $ad \neq 0$ there are $\ll 1 + \Delta_2|c|^{-1}C^2$ pairs of rationals satisfying (7.5.2) through (7.5.5).

We are now ready to bound $B_1(|bc| \geq D)$, which we denote B_2 for brevity. From the above, we see that

$$B_2 \ll \sum_{1\leq|c|\leq\Delta_1C^2} \sum_{|b|\neq 0} \sum_{\substack{|c|X<|a| \\ |d|\leq 25|c|X \\ ad=bc+1}} (1 + \Delta_2|c|^{-1}C^2),$$

where $X = A/5C$. Note that $X \gg 1$ because of (7.5.6). We break the range $1 \leq |c| \leq \Delta_1C^2$ into subranges of the form $Y < |c| \leq 2Y$, and we find that

$$\begin{aligned}
B_2 &\ll \sum_{Y} \sum_{Y<|c|\leq 2Y} \sum_{\substack{XY<|a| \\ |d|\leq 50XY \\ c|(ad-1)}} (1 + \Delta_2|c|^{-1}C^2). \\
&\ll \sum_{Y} \sum_{Y<|c|\leq 2Y} (1 + \Delta_2|c|^{-1}C^2) \sum_{XY<|a|\leq 50XY} \sum_{\substack{XY<|d|\leq 50XY \\ ad\equiv 1 \bmod c}} 1 \\
&\ll \sum_{Y} \sum_{Y<|c|\leq 2Y} (X^2Y^2|c|^{-1})(1 + \Delta_2|c|^{-1}C^2) \\
&\ll \Delta_1(\Delta_1 + \Delta_2)A^2C^2.
\end{aligned}$$

This completes the proof.

7.6 SEMICUBICAL POWERS OF INTEGERS

Here we obtain an upper bound for the number of integers $h_1, \ldots, h_4, k_1, \ldots k_4$ which satisfy

$$h_1^2 + \ldots + h_4^2 = k_1^2 + \ldots + k_4^2, \tag{7.6.1}$$

$$h_1 + \ldots + h_4 = k_1 + \ldots + k_4, \tag{7.6.2}$$

$$h_1^{3/2} + \ldots + h_4^{3/2} = k_1^{3/2} + \ldots + k_4^{3/2} + O(\delta H^{3/2}), \tag{7.6.3}$$

$$H < h_i, k_i \leq 2H. \tag{7.6.4}$$

LEMMA 7.19. *The number of solutions of*

$$\begin{aligned}
h_1^2 + h_2^2 + h_3^2 &= a \\
h_1 + h_2 + h_3 &= b
\end{aligned}$$

is at most $6d(6a-2b^2)$.

Proof. The equations imply that $x^2+3y^2=6a-2b^2$, where

$$x=h_1+h_2-2h_3 \quad \text{and} \quad y=h_1-h_2.$$

Now we use a result from the theory of quadratic forms, which states that the number of solutions of $x^2+3y^2=n$ is at most $6d(n)$. Since the equations

$$\begin{aligned} h_1+h_2-2h_3 &= x \\ h_1-h_2 &= y \\ h_1+h_2+h_3 &= b \end{aligned}$$

have at most one solution in integers h_1,h_2,h_3, the result follows.

LEMMA 7.20. *The solutions of (7.6.1) and (7.6.2) form families*

$$(h_1+t,\dots,h_4+t,k_1+t,\dots,k_4+t)$$

as t *varies. The number of families that contain a point with all coordinates in* $[H,2H]$ *and have*

$$|(k_i-h_1)(k_i-h_2)(k_i-h_3)(k_i-h_4)|\le \Delta H^4$$

for $i=1,\dots,4$ *is*

$$\ll H^3+\Delta H^{4+\epsilon}.$$

Proof. If $(h_1,\dots,h_4,k_1,\dots,k_4)$ is a solution, then so is $(h_1+t,\dots,h_4+t,k_1+t,\dots k_4+t)$. The product

$$(k_i-h_1)\dots(k_i-h_4)$$

is an invariant of the family. If all products of this type are zero, then each k_i is equal to an h_j. There are $\ll H^4$ integer octuplets with this property, and they form $\ll H^3$ families. Suppose, on the other hand, that the product involving k_1 is non-zero. Then the number of choices for $|k_1-h_1|,\dots,|k_1-h_4|$ is

$$\ll \sum_{n\le \Delta H^4} d_4(n)\ll \Delta H^{4+\epsilon}.$$

If we fix k_1, then we have $\ll \Delta H^{4+\epsilon}$ choices for $h_1,\dots,h_4$. After $k_1,h_1,\dots,h_4$ are chosen, the values of $k_2^2+k_2^2+k_3^2$ and $k_2+k_2+k_3$ are determined. Lemma 7.19 gives $\ll H^\epsilon$ choices for k_2,k_3,k_4. After redefining ϵ, we obtain the result of the lemma.

LEMMA 7.21. *The inequalities (7.6.1), (7.6.2), (7.6.3), and (7.6.4) have*

$$\ll H^{4+\epsilon}+\delta H^{5+\epsilon}$$

solutions.

Proof. The solutions form families of the type considered in Lemma 7.20. In considering these families, it is convenient to introduce the notation

$$D_n(x,y) \stackrel{\text{def}}{=} x_1^n + \ldots + x_4^n - y_1^n - \ldots - y_4^n.$$

The equation

$$D_{3/2}(h_1+t,\ldots,h_4+t,k_1+t,\ldots,k_4+t) = c$$

can be written as a polynomial equation in t of degree at most 767. Thus the solutions of the inequality

$$|D_{3/2}(h_1+t,\ldots,h_4+t,k_1+t,\ldots,k_4+t)| \le \delta H^{3/2}$$

form at most 768 intervals on the real line. By Lemma 7.20, those families in which each interval contains at most one integer solution contribute $\ll H^{4+\epsilon}$ to the total number of solutions. Suppose, on the other hand, that one such interval contains $T+1$ consecutive integers. By the Mean Value Theorem

$$|TD_{1/2}(h_1+t,\ldots,h_4+t,k_1+t,\ldots,k_4+t)| \le \frac{4}{3}\delta H^{3/2} \tag{7.6.5}$$

for some t within the interval. Write $u_i = \sqrt{h_i+t}$ and $v_i = \sqrt{k_i+t}$. We claim that for $n = 1,\ldots,4$ and $\Delta = \delta H/T$,

$$u_1^n + \ldots + u_4^n = v_1^n + \ldots + v_4^n + O(\Delta H^{n/2}). \tag{7.6.6}$$

For $n = 1$, (7.6.6) is a consequence of (7.6.5). For $n = 2$ and $n = 4$, (7.6.6) follows from (7.6.1) and (7.6.2). For $n = 3$, we note that if $s = [t]$, then $(h_1+s,\ldots,k_4+s)$ satisfies (7.6.3), and

$$\begin{aligned} &D_{3/2}(h_1+t,\ldots,h_4+t,k_1+t,\ldots,k_4+t) \\ =&D_{3/2}(h_1+s,\ldots,h_4+s,k_1+s,\ldots,k_4+s) + O(H^{1/2}). \end{aligned}$$

Now any symmetric function (with degree $d \le 4$) of $v_1,\ldots,v_4$ equals the same symmetric function of $u_1,\ldots,u_4$ to within $O(\Delta H^{d/2})$. Thus for $i = 1,\ldots,4$,

$$\prod_{j=1}^{4}(v_i - u_j) = \prod_{j=1}^{4}(v_i - v_j) + O(\Delta H^{d/2}) = O(\Delta H^{d/2}),$$

and

$$\prod_{j=1}^{4}(k_i - h_j) \ll \Delta H^2 \prod_{j=1}^{4}(v_i + u_j) \ll \Delta H^4.$$

By Lemma 7.20 there are

$$\ll H^{3+\epsilon} + \delta H^{5+\epsilon}2^{-r}$$

families in which some interval of solutions has $T \ge 2^r$. Multiplying by 2^r and summing gives $\ll H^{4+\epsilon} + \delta H^{5+\epsilon}$ solutions (when we redefine ϵ) from families with nontrivial intervals of solutions.

7.7 PROOF OF THE THEOREM

At several points in this section, we will use the notation $\sum_{a \approx A}$ to denote a sum over the range $c_1 A < a \le c_2 A$ for some unspecified constants c_1 and c_2.

Assume that $M_1 \le 2M$, and define

$$S = \sum_{M < m \le M_1} e(f(m)).$$

We begin with a variation of the Weyl-van der Corput method. Let N be some parameter to be chosen later. Assume for now that

$$N \le \frac{1}{2}(M_1 - M).$$

If $N < n \le 2N$ and $M_2 = M_1 - 2N$ then

$$S = \sum_{M < m \le M_2} e(f(m+n)) + O(N). \tag{7.7.1}$$

After averaging over n, we see that

$$S \ll \frac{1}{N} \sum_{M < m \le M_2} \Big| \sum_{N < n \le 2N} e(f(m+n) - f(m)) \Big| + N.$$

Let us write

$$f(m+n) - f(m) = g_m(n) + h_m(n),$$

where

$$g_m(n) = f'(m)n + \frac{1}{2} f^{(2)}(m) n^2 + \frac{1}{6} f^{(3)}(m) n^3.$$

We apply Lemma 7.2 with $a_n = e(g_m(n))$ and $\omega(n) = e(h_m(n))$. Clearly $|\omega(n)| \le 1$. Note also that

$$h'_m(n) = \int_0^n \int_0^u \int_0^v f^{(4)}(v+m)\, dw dv du,$$

so $|\omega'(n)| \ll FN^3 M^{-4}$. Therefore

$$S \ll N^{-1} \Big\{ \sum_{M < m \le M_2} \Big| \sum_{N < n \le 2N} e(g_m(n)) \Big| + \frac{FN^3}{M^4} \int_N^{N_1} \Big| \sum_{N < n \le x} e(g_m(n)) \Big| dx \Big\} + N.$$

If we assume that $FN^4 M^{-4} \le 1$, i.e.

$$N \le F^{-1/4} M \tag{7.7.2}$$

then it follows that there is some N_0 with $N_0 \leq 2N$ and

$$S \ll N^{-1}\Big\{ \sum_{M<m\leq M_2} \Big| \sum_{N<n\leq N_0} e(f'(m)n + \frac{1}{2}f^{(2)}(m)n^2 + \frac{1}{6}f^{(3)}(m)n^3)\Big|\Big\} + N.$$

By a theorem of Dirichlet, each middle coefficient $\frac{1}{2}f^{(2)}(m)$ has a rational approximation

$$\left|\frac{1}{2}f^{(2)}(m) + \frac{a}{c}\right| \leq \frac{1}{cN} \tag{7.7.3}$$

with $1 \leq c \leq N$ and $(a,c) = 1$. Let $m(x)$ be the solution of

$$-\frac{1}{2}f^{(2)}(m(x)) = x.$$

Then each m satisfying (7.7.3) may be written as

$$m = [m(a/c)] + l,$$

with $|l| \leq L(c)$ and $L(c) \approx (cFN)^{-1}M^3$. It will be convenient to have $L(c) \gg 1$, and we can insure this by assuming that

$$N \leq F^{-1/2}M^{3/2}. \tag{7.7.4}$$

We therefore see that S is majorized by

$$\frac{1}{N}\sum_{c\leq N}\sum_{\substack{a\approx cFM^{-2}\\(a,c)=1}}\sum_{|l|\leq L(c)}\Big|\sum_{\substack{N<n\leq N_0\\ m=m(a/c)+l\\ M\leq m\leq M_2}} e\left(f'(m)n + \frac{1}{2}f^{(2)}(m)n^2 + \frac{1}{6}f^{(3)}(m)n^3\right)\Big|$$
$$+ N.$$

The terms pertaining to the fractions a/c with $1 \leq c \leq C_0$ trivially contribute

$$\ll N^{-1}\sum_{1\leq c\leq C_0} L(c) \sum_{a\approx cFM^{-2}} N \ll C_0MN^{-1}.$$

Now assume that $C_0 < c \leq N$. Denote

$$b = b(a,c) = [cf'([m(a/c)])]$$

and

$$\mu = \mu(a/c) = \frac{1}{6}f^{(3)}(m(a/c)).$$

We claim that

$$\frac{1}{6}f^{(3)}(m) = \mu + O((cMN)^{-1})$$

and that

$$f'(m) = \frac{b-2al}{c} + O\left(\frac{1}{c} + \frac{M^3}{c^2 F N^2}\right). \tag{7.7.5}$$

To justify the first assertion, we observe that

$$\frac{1}{6} f^{(3)}(m) - \mu = \int_{m(a/c)}^{m} \frac{1}{6} f^{(4)}(u)\, du \ll \frac{L(c)F}{M^4} \ll \frac{1}{cMN}.$$

For the second assertion, we note that

$$\begin{aligned} f'(m) &= f'([m(a/c)]) + \int_{[m(a/c)]}^{m} f^{(2)}(u)\, du \\ &= f'([m(a/c)]) - \frac{2a}{c} l + \int_{[m(a/c)]}^{m} \left(f^{(2)}(u) - f^{(2)}(m(a/c))\right) du. \end{aligned}$$

The integrand is $\ll lFM^{-3}$, so the integral is $\ll l^2 F N^{-3}$. Moreover,

$$f'([m(a/c)]) = \frac{1}{c}[cf'([m(a/c)])] + O\left(\frac{1}{c}\right) = \frac{b}{c} + O\left(\frac{1}{c}\right).$$

We assume henceforth that

$$C_0 = F^{-1} M^3 N^{-2};$$

this allows us to rewrite (7.7.5) as

$$f'(m) = \frac{b-2al}{c} + O\left(\frac{1}{c}\right).$$

It follows that

$$f'(m) + \frac{1}{2} f^{(2)}(m) n + \frac{1}{6} f^{(3)}(m) n^2 = \frac{b-2al}{c} - \frac{a}{c} n + \mu n^2 + O\left(\frac{1}{c}\right).$$

By Lemma 7.2 we conclude that there is some $N_1 \le 2N$ independent of the variables of summation such that

$$S \ll \sum_{C_0 < c \le N} c^{-1} \sum_{a \approx cFM^{-2}} \sum_{\substack{|l| \le L(c) \\ \theta(c,a,l)}} \left| \sum_{N < n \le N_1} e\left(\frac{b-2al}{c} n - \frac{a}{c} n^2 + \mu n^3\right) \right| + \\ + N + C_0 M N^{-1};$$

where we have written $\theta(c, a, l)$ as shorthand for the condition $M < m(a/c) + l \le M_2$. It follows that for some A, C, L with

$$C_0 \le C \le N, A \approx CFM^{-2}, L \approx (CFN)^{-1} M^3,$$

we have

$$S \ll \frac{\log N}{C} \sum_{C<c\leq 2C} \sum_{\substack{A<a\leq 2A \\ (a,c)=1}} \sum_{\substack{|l|\leq L \\ \theta(c,a,l)}} |S(\mu; -a, b-2al; c)| + N + F^{-1}M^4N^{-3}.$$

Now we use the estimate for $S(\mu; a, b; c)$ derived in Lemma 7.16. The error terms contribute

$$\begin{aligned}&\ll C^{-1}\sum_a\sum_c\sum_l(N^{1/2}+\mu^{-1}N^{-2})(\log N)^2 \\ &\ll C^{-1}ACL(N^{1/2}+F^{-1}M^3N^{-2})(\log N)^2 \\ &\ll (MN^{-1/2}+F^{-1}M^4N^{-3})(\log N)^2\end{aligned}$$

The main term $\sum_h$ can be simplified a bit. First of all, we may use Lemma 7.2 to remove the factor $(3\mu ch)^{-1/4}$. Then we may replace the constraint $3\mu cN^2 < h < 3\mu cN_1^2$ by the weaker condition $h \approx H$ with

$$H = CFM^{-3}N^2 \qquad\qquad (1 < H < C)$$

by Lemma 7.3. The resulting inequality is

$$\begin{aligned}S \ll & C^{-3/2}F^{-1/2}M^{3/2}N^{-1/2}(\log N)^2 \cdot \\ &\cdot \sum_{c\approx C}\sum_{\substack{a\approx A \\ (a,c)=1}}\sum_{\substack{|l|\leq L \\ \theta(c,a,l)}}\Big|\sum_{h\approx H} e\left(\frac{\bar{a}}{c}\frac{h^2}{4}+\frac{\bar{a}b-2l+c\eta}{c}\frac{h}{2}-\frac{\nu h^{3/2}}{c}\right)\Big| \\ &+(N+MN^{-1/2}+F^{-1}M^4N^{-3})(\log N)^2,\end{aligned}$$

where

$$\nu = \nu(a,c) = \frac{4}{3}\left(2cf^{(3)}(m(a/c))\right)^{-1/2} \approx C^{-1/2}F^{-1/2}M^{3/2},$$

and η is a real number which does not depend on the variables of summation. (This term arises from the use of Lemma 7.3 and from the factor w^h.) By Hölder's inequality, we see that

$$S \ll \left\{C^{-3/4}F^{-1/2}M^{9/4}N^{-5/4}T^{1/4} + N + MN^{-1/2} + F^{-1}M^4N^{-3}\right\}(\log N)^2,$$

where T is defined as

$$T = \sum_{c\approx C}\sum_{\substack{a\approx A \\ (a,c)=1}}\sum_{\substack{|l|\leq L \\ \theta(c,a,l)}}\Big|\sum_{h\approx H} e\left(\frac{\bar{a}}{c}\frac{h^2}{4}+\frac{\bar{a}b-2l+c\eta}{c}\frac{h}{2}-\frac{\nu h^{3/2}}{c}\right)\Big|^4.$$

The sum T is a bilinear form of the type $\mathbf{B}(a, b; \mathbf{X}, \mathbf{Y})$ considered in Lemma 7.5, where $a(\mathbf{x})$ is the multiplicity of representations of $\mathbf{x}$ of the form

$$\mathbf{x} = (x_1, x_2, x_3) = \Big(\frac{1}{4}\sum_1^4(-1)^j h_j^2, \frac{1}{2}\sum_1^4(-1)^j h_j, \sum_1^4(-1)^j h_j^{3/2}\Big)$$

with relevant h_j, and $b(\mathbf{y})$ is the multiplicity of representations of the form

$$\mathbf{y} = (y_1, y_2, y_3) = (\frac{\bar{a}}{c}, \frac{\bar{a}b - 2l + c\eta}{c}, \frac{\nu}{c}).$$

Therefore

$$\mathbf{X} = (X_1, X_2, X_3,) \approx (H^2, H, H^{3/2})$$

and

$$\mathbf{Y} = (Y_1, Y_2, Y_3) = (1, 1, Y_3)$$

with

$$Y_3 \approx F^{-1/2}C^{-3/2}M^{3/2}.$$

By Lemma 7.5, we get

$$T^2 \ll H^{9/2}Y_3\mathbf{B}(b; \mathbf{X})\mathbf{B}(a; \mathbf{Y}).$$

In this context, $\mathbf{B}(b; \mathbf{X})$ is the number of pairs $\{(a, c, l), (a_1, c_1, l_1)\}$ such that

$$||\frac{\bar{a}}{c} - \frac{\bar{a}_1}{c_1}|| \ll H^{-2}, \tag{7.7.6}$$

$$||\frac{\bar{a}b + 2l}{c} - \frac{\bar{a}_1 b_1 + 2l_1}{c_1}|| \ll H^{-1}, \tag{7.7.7}$$

$$||\frac{\nu(a,c)}{c} - \frac{\nu(a_1,c_1)}{c_1}|| \ll H^{-3/2} \tag{7.7.8}$$

and $\mathbf{B}(a; \mathbf{Y})$ is the number of h_j's ,$1 \le j \le 8$, such that

$$\sum_1^4 (h_j^2 - h_{j+4}^2) = 0, \tag{7.7.9}$$

$$\sum_1^4 (h_j - h_{j+4}) = 0, \tag{7.7.10}$$

$$\sum_1^4 (h_j^{3/2} - h_{j+4}^{3/2}) \ll H^{1/2}. \tag{7.7.11}$$

From Lemma 7.21, we see that

$$\mathbf{B}(a; \mathbf{Y}) \ll H^{4+\epsilon}.$$

Now $LH \approx N$, so $LC^{-1} \approx NC^{-1} \gg H^{-1}$. From (7.7.7), we deduce that

$$||\frac{\bar{a}b}{c} - \frac{\bar{a}_1 b_1}{c_1}|| \ll \frac{L}{C}. \tag{7.7.12}$$

Moreover, for each set $(a, b, c; a_1, b_1, c_1)$ there are

$$\ll (C + L)L/H$$

pairs of (l, l_1) which satisfy (7.7.12). Therefore

$$\mathbf{B}(b; \mathbf{X}) \ll (C + L)LH^{-1}\mathbf{N}(A, C, H),$$

where $\mathbf{N}(A, C, H)$ is the number of quadruples $(a, c; a_1, c_1)$ which satisfy (7.7.6), (7.7.8), and (7.7.12).

To estimate $\mathbf{N}(A, C, H)$, we first need to transform condition (7.7.8). Let $h(x)$ be defined by the equation

$$h^3(x) = f^{(3)}(m(x)).$$

Then $h(x) \approx F^{1/3}M^{-1}$ and

$$h'(x) = \frac{f^{(4)}(m(x))}{3h^2(x)} \approx \left|\frac{h(x)}{x}\right|.$$

We now define

$$U = CF^{1/3}M^{-1},$$

so that $ch(a/c) \approx U$. For brevity, we write $\eta = ch(a/c)$ and $\eta_1 = c_1 h(a_1/c_1)$. With this notation, we have

$$||\frac{\nu(a, c)}{c} - \frac{\nu(a_1, c_1)}{c_1}|| \approx |\int_\eta^{\eta_1} u^{-5/2}\, du| \approx |\eta - \eta_1| U^{-5/2}.$$

It follows that if (7.7.8) is satisfied, then

$$|ch(a/c) - c_1 h(a_1/c_1)| \ll H^{-3/2}U^{5/2}.$$

We may therefore apply Lemma 7.18 with

$$\Delta_2 \approx H^{-3/2}U^{3/2} \approx F^{-1}M^3N^{-3}$$

and

$$\Delta_1 = H^{-2}$$

We are unable to take advantage of condition (7.7.7), so we simply discard it. Therefore

$$\mathbf{N}(A, C, H) \ll \Delta_1\Delta_2 A^2C^2 + AC + \Delta_1^2 A^2C^2 + \Delta_2 A^2 + \Delta_2 C^2.$$

Since $H \le C$, the fourth term is dominated by the first. Since $\Delta_2 = CL^{-1}\Delta_1$, we may rewrite the above as

$$\mathbf{N}(A, C, H) \ll \Delta_1 A^2C^2L^{-1}(C + L) + AC + \Delta_2 C^2.$$

We thereby obtain

$$\begin{aligned}\mathbf{B}(b;\mathbf{X}) &\ll (C+L)L\mathbf{N}(A,C,H)/H \\ &\ll CF^{-3}M^{11}N^{-10}+C^{-3}F^{-5}M^{17}N^{-12}+CF^{-1}M^{4}N^{-3}+ \\ &\quad +C^{-1}F^{-2}M^{7}N^{-4}+CF^{-3}M^{9}N^{-6}+C^{-1}F^{-4}M^{12}N^{-4}.\end{aligned}$$

In this last statement, we have used the inequality $(C+L)^2 \ll C^2+L^2$ to reduce the number of terms. We deduce that

$$\begin{aligned}T \ll &\{C^4F^{5/2}M^{-13/2}N^{7/2}+C^2F^{3/2}M^{-7/2}N^{5/2}+C^3F^3M^{-17/2}N^{13/2}+ \\ &+C^4F^{5/2}M^{-15/2}N^{11/2}+C^3F^2M^{-6}N^5\}M^{\epsilon}\end{aligned}$$

Plugging this into our earlier estimate of S yields

$$\begin{aligned}S \ll &\{F^{1/8}M^{5/8}N^{-1/8}+F^{1/8}M^{3/8}N^{3/8}+F^{3/8}M^{-1/4}N^{3/4}+ \\ &+F^{1/4}M^{1/8}N^{3/8}+M^{3/4}+N+MN^{-1/2}+F^{-1}M^4N^{-3}\}M^{\epsilon}.\end{aligned} \tag{7.7.13}$$

Up to now, we have been assuming that $N \le \frac{1}{2}(M_1-M)$. If we drop this assumption, then we still have (7.7.13) by virtue of the trivial estimate $S \ll N$. We now apply the Lemma 2.4 with the condition

$$N \le \min(F^{-1/4}M, F^{-1/2}M^{3/2}).$$

We find that

$$\begin{aligned}S \ll &\{F^{1/2}M^{-1/2}+F^{1/4}M^{1/4}+F^{3/16}M^{7/16}+F^{9/56}M^{1/2}++F^{1/9}M^{5/9} \\ &+F^{1/8}M^{9/16}+F^{1/14}M^{9/14}+F^{1/10}M^{3/5}+M^{7/9}+F^{-1/4}M\}M^{\epsilon}.\end{aligned}$$

This gives the desired estimate when $F^{3/7} \le M \le F^{4/7}$. The other cases were treated in Section 7.1, so this completes the proof.

7.8 NOTES

The material in this chapter has been adapted from the papers of Bombieri and Iwaniec (1986a,b) and Huxley and Watt (1988). As mentioned before, Watt (1989) has shown that

$$(89/560+\epsilon, 369/560+\epsilon)$$

is an exponent pair for any $\epsilon > 0$. The improvement comes form using Hölder's inequality with fifth powers rather than fourth powers, and from analogs of Lemmas 7.20 and 7.21 with 10 variables rather than 8.

APPENDIX

SOME AMAZING FUNCTIONS

In the late 1930's, A. Beurling observed that the entire function

$$B(z) = \frac{\sin^2 \pi z}{\pi^2} \left\{ \sum_{n=0}^{\infty} (z-n)^{-2} + \sum_{m=1}^{\infty} (z+m)^{-2} + 2z^{-1} \right\},$$

satisfies a simple and useful extremal property. Define

$$\text{sgn } (x) = \begin{cases} 1 & \text{if } x > 0; \\ 0 & \text{if } x = 0; \\ -1 & \text{if } x < 0. \end{cases}$$

Then

$$B(x) \geq \text{sgn } (x)$$

for all real x. The function $B(z)$ is entire of exponential type 2π. Beurling showed that if $F(z)$ is any entire function of exponential type 2π satisfying $F(x) \geq \text{sgn } (x)$ for all real x, then

$$\int_{-\infty}^{\infty} F(x) - \text{sgn } (x) dx \geq 1,$$

and equality occurs if and only if $F(z) = B(z)$.

In 1974, A. Selberg used the function B to give a simple proof of the large sieve inequality. Selberg noted that if $\chi_{\mathbf{I}}(x)$ is the characteristic function of the interval $\mathbf{I} = [a, b]$ and

$$C_{\mathbf{I}} = \frac{1}{2}\{B(b-z) + B(z-a)\}$$

then

$$C_{\mathbf{I}}(x) \geq \chi_{\mathbf{I}}(x)$$

for all real x. Moreover, the Fourier transform

$$\hat{C}_{\mathbf{I}}(t) = \int_{-\infty}^{\infty} C_{\mathbf{I}}(x) e(-tx) dx$$

vanishes outside the interval $[-1, 1]$. This property makes $C_{\mathbf{I}}$ particularly nice for applications.

In 1985, J. Vaaler showed how Beurling's function could be used to construct a trigonometric polynomial approximation to $\psi(x)$. For each positive integer N, Vaaler's construction yields a trigonometric polynomial ψ^* of degree N which satisfies

$$|\psi^*(x) - \psi(x)| \leq \frac{1}{2N+2} \sum_{|n| \leq N} \left(1 - \frac{|n|}{N+1}\right) e(nx). \tag{A.1}$$

This construction is useful for applications of exponential sums; see Chapter 4.

The material in this section is taken from Vaaler (1985). The functions discussed here satisfy a number of interesting extremal properties. Since we do not need these properties here, we do not give them; the interested reader can find them in Vaaler's article.

For complex z, define

$$H(z) = \frac{\sin^2 \pi z}{\pi^2}\left\{\sum_{n=-\infty}^{\infty} \frac{\text{sgn }(n)}{(z-n)^2} + \frac{2}{z}\right\},$$

$$J(z) = \frac{1}{2}H'(z),$$

$$K(z) = \frac{\sin^2 \pi z}{\pi^2 z}.$$

For real x, define

$$E(x) = H(x) - \text{sgn }(x).$$

LEMMA A.1. *If x is real then $|H(x)| \le 1$ and $|E(x)| \le K(x)$.*

Proof. Since H and E are odd functions, it suffices to consider $x > 0$. From the identity

$$\frac{\pi^2}{sin^2 \pi x} = \sum_{n=-\infty}^{\infty} \frac{1}{(x-n)^2},$$

we get

$$1 - H(x) = \frac{\sin^2 \pi x}{\pi^2}\left\{2\sum_{m=1}^{\infty} \frac{1}{(x+m)^2} + \frac{1}{x^2} - \frac{2}{x}\right\}.$$

To complete the proof, it suffices to show that

$$\frac{1}{x} - \frac{1}{2x^2} \le \sum_{m=1}^{\infty} \frac{1}{(x+m)^2} \le \frac{1}{x}. \tag{A.2}$$

Let $\{u\}$ denote the fractional part of u; i.e. $\{u\} = u - [u]$. Integration by parts shows that

$$\sum_{m=1}^{\infty} \frac{1}{(x+m)^2} = \int_0^{\infty} \frac{du}{(x+u)^2} - \int_0^{\infty} \frac{d\{u\}}{(x+u)^2} = \frac{1}{x} - 2\int_0^{\infty} \frac{\{u\}du}{(x+u)^3}. \tag{A.3}$$

Since $\{u\} \ge 0$, this proves the right hand inequality of (A.2). For the left-hand side, we note that $\{u\} = \psi(u) + 1/2$, so

$$2\int_0^{\infty} \frac{\{u\}du}{(x+u)^3} = \int_0^{\infty} \frac{du}{(x+u)^3} + 2\int_0^{\infty} \frac{\psi(u)du}{(x+u)^3} = \frac{1}{2x^2} + 2\int_0^{\infty} \frac{d\psi_1(u)}{(x+u)^3},$$

where

$$\psi_1(u) = \int_0^u \psi(t)\,dt = \frac{\{u\}(\{u\}-1)}{2}.$$

Integrating by parts again, we get

$$2\int_0^\infty \frac{\{u\}du}{(x+u)^3} = \frac{1}{2x^2} + \int_0^\infty \frac{6\psi_1(u)du}{(x+u)^4} \le \frac{1}{2x^2}$$

since $\psi_1(u) \le 0$ for all u. This together with (A.3) yields the left-hand side of (A.2).

LEMMA A.2. *The Fourier transform $\hat{J}(t)$ satisfies*

$$\hat{J}(t) = \begin{cases} 1 & \text{if } t = 0; \\ \pi t(1-|t|)\cot \pi t + |t| & \text{if } 0 < |t| < 1; \\ 0 & \text{if } t \ge 1. \end{cases}$$

Proof. Let

$$H_N(z) = \frac{\sin^2 \pi z}{\pi^2}\left\{\sum_{|m|\le N} \frac{\text{sgn}\,(m)}{(z-n)^2} + \frac{2}{z}\right\},$$

so that, as $N \to \infty$, $\lim H_N(z) = H(z)$ and $\lim H_N'(z) = 2J(z)$ as $N \to \infty$. From the identities

$$K(z) = \int_{-1}^1 (1-|t|)e(tz)dt \text{ and } zK(z) = \frac{1}{2\pi i}\int_{-1}^1 \text{sgn}\,(t)\,e(tz)dt,$$

we see that

$$\begin{aligned} H_N(z) &= \sum_{|m|\le N} \text{sgn}\,(m)K(z-m) + 2zK(z) \\ &= \int_{-1}^1 \Big\{(1-|t|)\sum_{|m|\le N} \text{sgn}\,(m)e(-mt) + \frac{1}{\pi i}\text{sgn}\,(t)\Big\}e(tz)\,dt. \end{aligned}$$

Apply the differential operator $\frac{1}{2}\frac{d}{dz}$ to both sides to get

$$\frac{1}{2}H_N'(z) = \int_{-1}^1 \{\pi i t(1-|t|)\sum_{|m|\le N} \text{sgn}\,(m)e(-mt) + \text{sgn}\,(t)\}\,e(tz)\,dt.$$

Upon using the relation

$$\sum_{|m|\le N} \text{sgn}\,(m)e(-mt) = \frac{1}{i}\left(\cot \pi t - \frac{\cos \pi(2N-1)t}{\sin \pi t}\right)$$

we find that

$$\frac{1}{2}H_N'(z) = \int_{-1}^{1} \{\pi t(1-|t|)\cot \pi t + t\text{sgn }(t)\}\, e(tz)\, dt$$
$$+ \int_{-1}^{1} \frac{\pi t(1-|t|)}{\sin \pi t}\, cos(\pi(2N-1)t)e(tz)dt.$$

By the Riemann-Lebesgue lemma, the second integral tends to 0 as N tends to infinity. Therefore

$$J(z) = \int_{-1}^{1} \{\pi t(1-|t|)\cot \pi t + t\,\text{sgn }(t)\}e(tz)dt \tag{A.4}$$

Now let $\phi(t)$ be the function defined by $\phi(t) = \pi t(1-|t|)\cot \pi t + t$ for $-1 < t < 2$, $t \neq 0,1$, and is defined so that ϕ is continuous at $t=0$ and $t=1$. Equation (A.4) may be rewritten as

$$J(z) = 2\int_0^1 \phi(t) cos(2\pi tz)dt.$$

After integrating by parts three times, we obtain

$$J(z) = \frac{1}{(2\pi z)^3} \left\{2\int_0^1 \phi'''(t)\sin(2\pi tz)dt - \frac{4\pi^2}{3}\sin(2\pi z)\right\}.$$

It follows that when x is real, $J(x) \ll (1+|x|)^{-3}$. Therefore, J is integrable on $\mathbf{R}$, and the desired formula for $\hat{J}$ follows from (A.4) and the Fourier inversion formula.

COROLLARY A.3. *The Fourier transform of* $E(x) = H(x) - \text{sgn }(x)$ *is*

$$\hat{E}(t) = \begin{cases} 0 & \text{if } t = 0; \\ (\hat{J}(t)-1)(\pi i t)^{-1} & \text{if } t \neq 0. \end{cases}$$

Proof. Since E is odd, we see immediately that $\hat{E}(0) = 0$. Assume now that $t \neq 0$. An integration by parts shows that

$$\hat{E}(t) = \frac{1}{2\pi i t}\int_{-\infty}^{\infty} e(-tx)\, dE(x).$$

Another integration by parts shows that

$$\frac{1}{2}\int_{-\infty}^{\infty} e(-tx)\, dE(x) = \hat{J}(t) - 1,$$

and the corollary follows.

THEOREM A.4. *Define $B(z) = H(z) + K(z)$ and $b(z) = H(z) - K(z)$. Let* **I** *be the interval $[\alpha, \beta]$, and let $\chi_{\mathbf{I}}$ be the characteristic function of I. Finally, define*

$$C_{\mathbf{I}}(z) = \frac{1}{2}\{B(\beta - z) + B(z - \alpha)\}, \; c_{\mathbf{I}}(z) = \frac{1}{2}\{b(\beta - z) + b(z - \alpha)\}.$$

(a) *If x is real then $b(x) \le \operatorname{sgn}(x) \le B(x)$.*

(b) *If x is real then $c_{\mathbf{I}}(x) \le \chi_{\mathbf{I}}(x) \le C_{\mathbf{I}}(x)$.*

(c) *$\hat{c}_{\mathbf{I}}(0) = \beta - \alpha - 1$ and $\hat{C}_{\mathbf{I}}(0) = \beta - \alpha + 1$.*

(d) *If $|t| \ge 1$ then $\hat{c}_{\mathbf{I}}(t) = 0$ and $\hat{C}_{\mathbf{I}}(t) = 0$.*

Proof. Part (a) follows immediately from Lemma A.1. For $x \neq \alpha$ and $x \neq \beta$ we have

$$\chi_{\mathbf{I}}(x) = \frac{1}{2}\{\operatorname{sgn}(\beta - x) + \operatorname{sgn}(x - \alpha)\},$$

so (b) follows immediately from (a). Since $C_{\mathbf{I}}$ and $c_{\mathbf{I}}$ are continuous, the restrictions on x may be removed.

From the above representation for $\chi_{\mathbf{I}}$ we see that

$$\hat{C}_{\mathbf{I}}(t) = \frac{1}{2}\{\hat{E}(-t)e(-\beta t) + \hat{E}(t)e(-\alpha t) + \hat{K}(-t)e(-\beta t) + \hat{K}(t)e(-\alpha t)\} + \hat{\chi}_{\mathbf{I}}(t).$$

From Corollary A.3, we see that

$$\hat{C}_{\mathbf{I}}(0) = \hat{K}(0) + \hat{\chi}_{\mathbf{I}}(0) = \beta - \alpha + 1.$$

Moreover, if $|t| \ge 1$ then

$$\frac{e(-\alpha t) - e(-\beta t)}{2\pi i t} + \hat{\chi}_{\mathbf{I}}(t) = 0.$$

This proves the claimed results for $C_{\mathbf{I}}$; the results for $c_{\mathbf{I}}$ are proved in a similar manner.

THEOREM A.5. *Let N and T be positive real numbers, and let a_n be a sequence of complex numbers with $a_n = 0$ if $n \le N$ or $n > 2N$. Suppose g is a real-valued function with $|g(m) - g(n)| > \delta$ whenever $N < m, n \le 2N$ and $m \neq n$. Then*

$$(T - \delta^{-1}) \sum_n |a_n|^2 \le \int_T^{2T} |\sum_n a_n e(tg(n)|^2 dt \le (T + \delta^{-1}) \sum_n |a_n|^2.$$

Proof. Let $\mathbf{I} = [\delta T, 2\delta T]$, and $F(x) = C_{\mathbf{I}}(\delta x)$. Then

$$F(x) \ge \begin{cases} 1 & \text{if } T \le x \le 2T, \\ 0 & \text{otherwise.} \end{cases}$$

We then find that

$$\begin{aligned}\int_T^{2T} |\sum_n a_n e(tg(n))|^2 dt &\le \int_{-\infty}^{\infty} F(t) |\sum_n a_n e(tg(n))|^2 dt \\ &= \sum_{m,n} a_m \overline{a_n} \hat{F}(g(m) - g(n)).\end{aligned} \tag{A.5}$$

From the definition of F we see that

$$\hat{F}(u) = \frac{1}{\delta} \hat{C}_{\mathbf{I}} \left(\frac{u}{\delta} \right).$$

If $N \le m, n \le 2N$ and $m \ne n$, then

$$\hat{F}(g(m) - g(n)) = 0.$$

Thus only the diagonal terms contribute to the sum in (A.5). Therefore

$$\int_T^{2T} |\sum_n a_n e(tg(n))|^2 \, dt \le \hat{F}(0) \sum_n |a_n|^2 = (T + \delta^{-1}) \sum_n |a_n|^2.$$

The same argument with $c_{\mathbf{I}}$ in place of $C_{\mathbf{I}}$ gives the lower bound

$$\int_T^{2T} |\sum_n a_n e(tg(n))|^2 dt \ge (T - \delta^{-1}) \sum_n |a(n)|^2.$$

In our final theorem, we will give Vaaler's construction of a trigonometric polynomial that majorizes $\psi(x)$. Here, we will use $F_\delta(x)$ to denote $\delta F(x/\delta)$. Note that if F is integrable then $\hat{F}_\delta(t) = \hat{F}(t/\delta)$.

THEOREM A.6. *The trigonometric polynomial*

$$\psi^*(x) = - \sum_{1 \le |n| \le N} (2\pi i n)^{-1} \hat{J}_{N+1}(n) e(nx)$$

satisfies

$$|\psi^*(x) - \psi(x)| \le \frac{1}{2N+2} \sum_{|n| \le N} \left(1 - \frac{|n|}{N+1} \right) e(nx).$$

Proof. By the Poisson summation formula and Corollary A.3,

$$\frac{1}{2N+2} \sum_m E_{N+1}(x+m) = \frac{1}{2\pi i n} - \sum_n (\hat{J}_{N+1}(-n) - 1) e(nx) = \psi^*(x) - \psi(x).$$

The last step uses the fact that $\hat{J}$ is even. Combining this with Lemma A.2, we get

$$|\psi^*(x) - \psi(x)| \le (2N+2)^{-1} \sum_m K_{N+1}(x+m) = (2N+2)^{-1} \sum_m \hat{K}_{N+1}(n) e(-nx).$$

Since

$$\hat{K}_{N+1}(n) = \hat{K} \left(\frac{n}{N+1} \right) = 1 - \frac{|n|}{N+1},$$

this completes the proof.

Bibliography

Works are listed alphabetically by author(s). Those by the same author(s) are listed chronologically.

Apostol, T.M. (1974). *Mathematical Analysis.* 2nd ed. Reading: Addison-Wesley.

Bombieri, E. and Iwaniec, H. (1986a). On the order of $\zeta(1/2+it)$, *Annali Scuola Normale Superiore - Pisa* (4) **13**, 449-472.

Bombieri, E. and Iwaniec, H. (1986b). Some mean-value theorems for exponential sums, *Annali Scuola Normale Superiore - Pisa* (4) **13**, 473-486.

Chen, J-R. (1963). The lattice points in a circle, *Sci. Sinica* **12** , 633-649.

Corput, J.G. van der (1921). Zahlentheoretische Abschätzungen, *Math. Ann.* **84**, 53-79.

Corput, J.G. van der (1922). Verscharfung der Abscha:tzungen beim Teilerproblem, *Math. Ann.* **87**, 39-65.

Corput, J.G. van der (1928). Zum Teilerproblem, *Math. Ann.* **98**, 697-716.

Corput, J.G. van der (1936). Über Weylsche Summen, *Mathematica B Leiden* **5**, 1-10.

Davenport, H. (1966). *Multiplicative Number Theory.* 1st edn. Chicago: Markham. 2nd ed. revised by Montgomery, H.L. (1980). Graduate Texts in Mathematics, 74. Berlin: Springer-Verlag.

Graham, S.W. and Kolesnik,G. (1987). One and two dimensional exponential sums *Analytic Number Theory and Diophantine Problems* (editors A.C. Adolphson, J.B. Conrey, A. Ghosh, and R.I. Yager, Birkhauser, Boston, 1987) 205-222.

Hardy, G.H. (1910). On certain definite integrals considered by Airy and by Stokes, *Quart. J. Math.*, **41**, 226-240.

Hardy, G.H. and Littlewood, J.E. (1921). The zeros of Riemann's zeta-function on the critical line, *Math. Zeit.* **10**, 283-317.

Hardy, G.H. and Wright, E.M. (1979). *An Introduction to the theory of numbers.* 5th ed. Oxford: Clarendon Press.

Haneke, W. (1963). Verschärfung der Abschätzung von $\zeta(1/2+it)$, *Acta Arith.* **8**, 357-430.

Heath-Brown, D.R. (1983). The Pjateckiĭ-Šapiro prime number theorem, *J. No. Theory* **16**, 242-266.

Heath-Brown, D.R. and Iwaniec, H. (1979). On the difference between consecutive primes, *Invent. Math.* **55**, 49-69.

Herzog,F. and Piranian,G. (1949). Sets of convergence of Taylor Series I, *Duke Math. Jnl.* **16**, 529-534.

Hua, L.K. (1941). The lattice points in a circle, *Quart. J. Math.* (Oxford) **12**, 193-200.

Huxley, M.N. and Watt, N. (1988). Exponential sums and the Riemann zeta-function, *Proc. London Math. Soc.* (3) **57**, 1-24.

Ivić, A. (1985) *The Riemann-zeta function*, Wiley-Interscience: New York.

Iwaniec, H. and Mozzochi, C.J. (1988). On the divisor and circle problems, *Journal of Number Theory* **29** 60-93.

Kolesnik, G. (1967). The distribution of prime numbers in sequences of the form $[n^c]$ (Russian), *Mat. Zametki* **2**, 117-128.

Kolesnik, G. (1969). Improvement of remainder term for the divisors problem (Russian), *Mat. Zametki* **6**, 545-554.

Kolesnik, G. (1981). On the estimation of multiple exponential sums, *Recent Progress in Analytic Number Theory* Vol. 1 (editors H. Halberstam and C. Hooley, Academic Press, New York, 1981) 247-256.

Kolesnik, G. (1981). On the number of abelian groups of a given order, *J. Reine Angew. Math.* **329** 164-175.

Kolesnik, G. (1982). On the order of $\zeta(1/2+it)$ and $\Delta(R)$, *Pac. Jnl. of Math.* **98**, 107-122.

Kolesnik, G. (1985a). On the method of exponent pairs. *Acta Arith.* **45**, 115-143.

Kolesnik, G. (1985b). On a paper of D.-R. Heath-Brown. *Pacific J. Math.* **118**, 437-447.

Kusmin,R. (1927). Über einige trigonometrische Ungleichungen, *J. Math.-Phys.,Leningrad* **1** 233-239. (Russian)

Landau, E. (1928). Über einer trigonometrische Summen, *Nachr. Ges. Wiss. Göttingen* 21-24.

Min, S.H. (1949). On the order of $\zeta(1/2+it)$, *Trans. Amer. Math. Soc.* **65**, 448-472.

Montgomery,H.L. and Vaughan, R.C. (1981) On the distribution of squarefree numbers, *Recent Progress in Analytic Number Theory* (eds. H. Halberstam and C. Hooley) New York:Academic Press. Vol. 1, 247-256.

Mordell, J.L. (1958) On the Kusmin-Landau inequality for exponential sums, *Acta Arith.* **4**, 3-9.

Piatetski-Shapiro , I.I. (1953). On the distribution of prime numbers in sequences of the form $[f(n)]$, *Math. Sb.* **33**, 559-566.

Phillips, E. (1933). The zeta-function of Riemann; further developments of van der Corput's method, *Quart. J. Math.* (Oxford) **4**, 209-225.

Rankin, R.A. (1955). Van der Corput's method and the theory of exponent pairs, *Quart. J. Math.* (Oxford) (2) **6**, 147-153.

Richert, H.-E. (1953). Verscharfüng der Abschätzung beim Dirichletschen Teilerproblem, *Math. Zeit.* **58**, 204-218.

Schmidt, P.G. (1968). Zur Anzahl Abelscher Gruppen gegebner Ordnung I, *Acta Arith.* **13**, 405-417.

Srinivasan, B.R. (1963). The lattice point problem in many dimensional hyperboloids I, *Acta Arith.* **8**, (1963) 173-204.

Srinivasan, B.R. (1965). The lattice point problem in many dimensional hyperboloids III, *Math. Ann.* **160**, (1965) 280-311.

Srinivasan, B.R. (1973). On the number of abelian groups of a given order, *Acta Arith.* **23** 195-205.

Titchmarsh, E.C. (1931) On van der Corput's method and the zeta-function of Riemann II, *Quarterly Journal of Mathematics* (Oxford Series) **2** 313-320.

Titchmarsh, E.C. (1934a). On Epstein's zeta-function, *Proc. London Math. Soc.* (2) **36**, 485-500.

Titchmarsh, E.C. (1934b). The lattice points in a circle, *Proc. London Math. Soc.* (2) **38**, 96-155. See also Corrigendum, *op. cit.* (1935) 55.

Titchmarsh, E.C. (1942). On the order of $\zeta(1/2+it)$ *Quart. J. Math.* (Oxford) **13** 11-17.

Titchmarsh, E.C. (1951) *The theory of the Riemann-zeta function*, Oxford: Clarendon Press. 2nd ed. revised by Heath-Brown, D.R. (1986).

Vaaler, J.D. (1985). Some extremal problems in Fourier analysis, *Bull. Amer. Math. Soc.* (2) **12**, 183-216.

Vaughan, R.C. (1980). An elementary method in prime number theory, *Acta Arithmetica* **37**, 111-115.

Vinogradov, I.M. (1985). *Selected Works*, Springer: Berlin.

Voronoï, G. (1903). Sur un problème du calcul des fonctions asymptotiques, *J. Reine Angew. Math.* **126**, 241-282.

Walfisz, A. (1924). Zur Abschätzung von $\zeta(\frac{1}{2}+it)$, *Göttinger Nachrichten*, 155-158.

Watt, N. (1989). Exponential sums and the Riemann zeta-function II, *J. London Math. Soc. (2)* **39** (1989) 385-404.

Weyl, H. (1916). Über die Gleichverteilung von Zahlen mod. Eins *Math. Ann.* **77**, 313-352.

Weyl, H. (1921). Zur Abschätzung von $\zeta(1+ti)$, *Math. Zeit.* **10**, 88-101.

Index